BHB

Further Advances in Chemical Information

Further Advances in Chemical Information

Edited by

H. Collier
Infonortics Ltd., Calne, Wiltshire

ROYAL
SOCIETY OF
CHEMISTRY

The Proceedings of the Montreux International Chemical Information
Conference, held at Annecy, France, 18–20 October 1993

Special Publication No. 142

ISBN 0-85186-545-3

A catalogue record for this book is available from the British Library

Published by The Royal Society of Chemistry,
Thomas Graham House, Science Park, Cambridge
CB4 4WF

Printed and bound by Bookcraft (Bath) Ltd.

Preface

The 1993 International Chemical Information Conference that took place in Annecy, France in October 1993 had a number of interesting themes; themes that to some extent reflected the somewhat uncertain times that currently confront both the information industry and its customers in the chemical and pharmaceutical industry.

The concept of *change* was certainly high on the agenda. Robert Massie (Chemical Abstracts) outlined changes in CAS's attitudes and plans in the 1990s. Stephen Heller (USDA) gazed into the future of chemical information and made a number of predictions. Wolfgang Donner (Bayer) talked about changes in demand, and about the reasons for the demise of IDC. Monica Pronin (American Petroleum Institute) talked about survival in the 1990s.

A second theme in this meeting was the rapid developments in in-house systems, especially 2-D and 3-D searching, spreadsheets and QSAR. Updating presentations were given by Roger Upton (Chemical Design), David Weininger (Daylight), Tad Hurst (Tripos Associates), and Phil McHale (Molecular Design). John Brennan (European Patent Office) also contributed an examination of the advanced information retrieval systems being developed in the EPO. Two commercial competitors, Research Publications and MicroPatent, contrasted their respective abilities to deliver US patents to in-house chemical users.

The Annecy conference is also well known for its technical and research papers. This year, there were two interesting papers on aspects of neural networks (with Jure Zupan discussing counter-propagation learning theory, and David Elrod discussing chemical applications). Bernhard Rohde (Ciba) discussed reaction type informetrics, and Gérard Kaufmann (Strasbourg University) explained the GRAMS project for learning about synthetic methods.

In the information world, chemists are interesting, since they take their information seriously, and since they tend to be pioneers, roaming out on the leading edges of technology, demanding to extend searching and discovery to encompass new features and new domains. This current Recent Advances book again serves to spotlight what is happening, as well as to place a marker as to where we have reached during the period up to the end of 1993.

Harry Collier

Editor

Contents

Chemical Abstracts Service: vista for the 1990s

Robert J. Massie

Chemical Abstracts Service, Columbus, Ohio, USA

For those of you who may not know Chemical Abstracts Service well, I will begin with some basic background about our organisation.

The present-day CAS grew out of the publication Chemical Abstracts, which was established by the American Chemical Society some 86 years ago. Chemical Abstracts began as virtually a one-person enterprise. Today, CAS employs nearly 1,400 staff and has an annual operating budget in excess of $100 million.

CAS's mission is to serve the needs of users of chemical and related information, worldwide, by providing value-added pathways to primary scientific literature and other data. As part of this mission, CAS is the operator in North America of the STN International online network, which now provides access to about 160 databases on a wide range of scientific and technical topics.

As an integral part of the American Chemical Society, CAS also supports and participates in the Society's broader missions of advancing chemical science and chemical education. We work closely with other divisions of the Society, particularly the Publications Division, which publishes the Society's journals.

CAS is a self-supporting unit within ACS. We receive no subsidies to pay our expenses and must exist in the competitive marketplace. More than 99 percent of our revenue comes from third parties making day-in, day-out marketplace decisions to use and pay for our services. We do try to return a modest surplus to ACS for pursuit of its chartered purposes. This surplus has tended in recent years to be in the range of only two to three percent of revenue.

Changes in the 1990s

CAS today is refocusing itself and changing in many ways. Some very significant changes have already taken place so far in the 1990s. Let me review them with you.

In 1991, the American Chemical Society established a new governance structure for CAS: the seven-member governing board that includes information industry leaders and ACS members chosen for their management experience in industry and academe. The governing board functions as CAS's board of directors. The governing board is chaired by the Executive Director of ACS; the Chairman of the ACS Board of Directors and the Director of CAS are *ex officio* members.

Within CAS, in the course of the past year, we have named four new senior managers, all of whom have come to CAS from leading organisations. Our new Director of Editorial Operations, Dr James Lohr, comes to us with 27 years of research and management experience with the DuPont Company, including two years as General Director of the Technology Division of DuPont Japan. Our Marketing Director, Suzan Brown, was formerly a Marketing Director with Mead

Data Central with 15 years of experience in the information industry. Our Finance Director, Peter Roche, was previously Vice President, Finance, of Honda Manufacturing of America. Our Director of Human Resources, Tom Murrill, was a Vice President at Burlington Airline Express and Chemlawn Services Corporation.

Together with several key veteran CAS directors, the new management team is committed to four main thrusts:

- Focusing CAS outward on its customers and on being a first-rate service organisation

- Controlling costs to bring users products and services at the best possible value points — without sacrificing the integrity or quality of our databases

- Aggressively pursuing development of new products and services in close collaboration with our user community

- Making CAS a positive, team-oriented place to work to better fulfil our mission of meeting users' needs for scientific and technical information.

Focus on customers

Focus on customers means many things. We will consult customers worldwide on product and policy decisions through personal contacts and surveys and through the establishment of so-called 'topic advisory groups' to provide frequent input from users with specific interests. The first of these topic advisory groups, which are essentially an expansion of our former geographically based 'User Councils,' are now being formed.

In North America, we have expanded our sales force to 14 individuals based in various regions of the country so that all of our major North American customers have frequent personal contact with someone from CAS. We have also expanded our North American customer service and help desk staff and hours.

Of course, a large portion of our customers are in Europe, and we wish to serve those customers better. Our European customers have already been seeing, and will continue to see, much more of CAS senior management in Europe. You will also be seeing much more of CAS technical service staff in Europe in the immediate future. Ultimately, I believe that ACS and CAS will determine that we need a sustained European presence, perhaps through an ACS or CAS office in one or more cities in Europe.

New pricing policies

We recognise that in the current economic climate our customers and users are under great pressure to control and reduce costs. We must respond to this pressure and help our customers succeed. Our decisions over the past year reflect this commitment. Since I arrived at CAS in mid-1992, there has been no increase in prices of CAS files on STN. We also reduced the 1993 prices in deutschmarks, francs and pounds, and announced that they would hold for at least 1993. I think these actions are clear indications that we are serious about working to meet our customers' needs in these difficult economic times.

Moreover, with the approval of the CAS Governing Board, I have publicly committed CAS to a moderate pricing policy. We will in the future seek to impose only moderate price increases, and only as absolutely necessary to meet our costs, and, as I mentioned earlier, we are working hard to control those costs. In fact, we have instituted an across-the-board 'cost effectiveness' review at CAS. All aspects of our operations will be carefully monitored to ensure that we meet the highest efficiency levels of information industry companies of comparable size.

We also are constantly re-evaluating our pricing methodology. How to most appropriately charge for electronic information services continues to be perhaps the most vexing issue in the information industry today. Both online service operators and database producers are seeking alternatives to the traditional connect-time pricing of online services. Search-term charges, which have been CAS's alternative to time-based charging, represent one alternative, but they have been unpopular with some users. Since July of this year, we have offered users of CAS bibliographic files on STN International the option of paying low connect-hour charges with search-term charges or a higher hourly rate without search-term charges. We are not eliminating search-term pricing; we believe it is the right approach for some search protocols. But we are offering the alternative of connect-time pricing for those who want it.

Longer term, we will work towards flat rate or subscription prices. We believe that to bring end-user researchers to online searching we need to eliminate metered pricing.

New and enhanced products and services

In late 1992, we established a dedicated, senior-level, new product development team under the direction of Dr Rudolph Potenzone. We have recently begun to see the fruits of its work.

One area in which you will be seeing an array of new products from CAS is CD-ROM. CAS launched its first CD-ROM product only last year, the Chemical Abstracts 12th Collective Index on CD-ROM. A companion product, the 12th Collective CA Abstracts on CD-ROM, was introduced in mid-1993.

At the ACS national meeting in August, we announced a new family of CD-ROM products, CASurveyor. Each disc in this series will bring together approximately three years of information from Chemical Abstracts in a broad area of research — chromatography, magnetic resonance, food and feed chemistry, organometallic chemistry and cancer chemical research are the initial topics in the series.

Information on the discs can be accessed by title or text words, controlled index terms, chemical substance names, CAS Registry Numbers, molecular formulae, or author or inventor names through a simple, user-friendly interface. Chemical structure diagrams are displayed along with textual information. We will be expanding this series in future years, and we continue to look at other possible CD-ROM products.

One of the significant trends in information is the growing availability of full-text and graphics of patents and journals in electronic form both online and through CD-ROM or other distribution media. This is another area in which our new product development team is currently working.

The Chemical Journals Online group of files on STN International represents the oldest and largest experiment in online delivery of full text of scientific journals. This group of files includes the full text of 48 journals published by the American Chemical Society, Elsevier, John Wiley and Sons, the Royal Society of Chemistry, VCH Verlagsgesellschaft, and the Association of Official Analytical Chemists.

These files currently contain only the text of the journals, but an experiment currently underway at Cornell University is providing the prototype of a future service. The Chemistry Online Retrieval Experiment (CORE), a collaborative effort of Cornell University, the ACS Publications Division, CAS, the Online Computer Library Center and Bell Communications Research, enables users at Cornell to retrieve articles from 20 ACS journals, complete with illustrations, tables and mathematical formulae, through a local area network by full-text searching, searching controlled indexing provided by CAS, or by browsing tables of contents of journals. Users can switch to articles on related topics or to referenced articles using hypertext-type links and have the choice of displaying bitmapped page images that look like the original journal pages or 'reconstructed' pages that combine text with separate, scanned graphic images. The CORE project is evaluating what kind of interface best meets users' needs and which form of display is preferred.

A service of the not-too-distant future is likely to combine online searching of a secondary database to identify citations or references of interest with programmed retrieval and display of full page images of the cited articles or patents on the searcher's workstation. Such images may be delivered online or retrieved automatically from local optical storage. How many journals are accessible in this manner will, of course, depend on the journal publishers to a large degree. I can assure you that the American Chemical Society and CAS will be active participants in this emerging field. CAS is working closely with the ACS Publications Division in this area, and we anticipate announcing our plans in greater detail by year end.

We also continue to enhance and expand our existing services and databases. The STN International online service, which marks its 10th anniversary this year, will be 're-introduced' in December with a number of important new databases, including the Derwent World Patent Index file, and some significant new features. Patent graphics will be available in the World Patent Index file, and STN Express personal computer software will provide enhanced displays and prints of images. STN in the future will be more focused — on key database clusters in chemistry, pharmaceuticals, energy, health and safety and patents. We will emphasise improved customer service and tighter control of operating costs, and we will set higher technical standards for performance and availability to meet users' expectations.

Over the past several years we have added protein and nucleic acid sequence data to our Registry file of chemical substance information, which now contains the largest collection of biosequence data available online. We have also introduced stereochemical display in Registry, and stereochemical searching will be available in December. Further Registry enhancements will continue to focus on stereochemistry, and we continue to look for ways to improve our handling of substance data and the content of Registry.

We also have expanded the coverage of and added search features to our CASRE-ACT database of chemical reactions, have added abstracts and CAS Registry Numbers to our CApreviews current-awareness database, and continued to enhance our MARPAT database of Markush structures from chemical patents. We have acquired the CHEMLIST database on regulated chemicals from the American Petroleum Institute and substantially expanded its content.

Internally, we continue to work on improving the currency of our databases and improving efficiency and reducing costs of document analysis wherever we can without compromising quality. A new workstation-based system installed over the past several years has enhanced the editorial environment and increased the amount of computer support provided to our document analysts and the amount of direct online input of data. Computer generation of CA index names for chemical substances is being put into place this year, and we are looking at ways in which we might make this capability publicly available.

Towards the 21st century

Looking toward the remainder of the 1990s, we see a number of technological and market trends to which all information providers are going to have to respond:

- Use of CD-ROM is expanding rapidly.

- Cheaper and faster computers combined with graphical user interfaces are providing significant improvement in desktop tools available to individual researchers.

- Users want to be able to access public information together with proprietary information.

- Increasing network speeds are allowing distribution of more complex data, including full-page images with text and graphics.

- 'Expert systems' will increasingly be used to aid information processing.

CAS's main objective for the remainder of the 1990s is to get information into the hands of those who need it — by whatever delivery method or medium best suits the users' needs, at a fair price, and with access tools that are convenient, easy to use and state-of-the-art. In doing this, we will ensure that we provide 'personalised' services for individual users — our new CD-ROM services are a step in that direction. We will provide better, simpler and more graphic interfaces with our electronic databases for both end-user searchers and information specialists — we will have something to announce in that regard in early 1994. We will explore new approaches to searching and information retrieval, including hypertext navigation, concept searching and similarity searching. We will work closely with client organisations on information retrieval systems that best meet their specific needs, and we will work with and co-operate with other information providers and others in the information industry as necessary and appropriate to meet our users' needs. Our long-term goal is to prepare CAS to be the central source of chemical and related scientific information in the 21st century, as it has been for most of this century.

A renewed commitment

To be Director of CAS is to be the steward and guardian of an important international resource, and I feel the responsibility keenly. I am immensely optimistic about the future of CAS and about the opportunities that new and developing technologies afford for us to perform our mission more effectively. We have new governance and direction and a renewed commitment to meet the needs of users of chemical and related information, to work closely with client organisations, to include customers and users in our product and policy decisions, to control our costs, and to price fairly. We will act responsibly and vigorously to fulfil the public mission set out in the ACS Charter: "to encourage in the broadest and most liberal manner the advancement of chemistry in all its branches."

We appreciate your continued interest in and support of CAS and seek your suggestions on how we might better serve you.

The chemical information flow into and out of Japan

Hideaki Chihara

*Japan Association for International Chemical Information, Gakkai Center Bldg.
2-4-16 Yayoi, Bunkyoku, Tokyo 113, Japan*

I. Introduction

The lack of availability of Japanese information has been a subject of concern to the US Congress and at international meetings for some time [1]. As a public hearing in the US Congress revealed, there are no legislative restrictions which prevent the flow of information from Japan to the rest of the world. Nevertheless the quantity of Japanese information out of all the information used in the world is much less than one per cent. The reasons why people feel that Japanese information is difficult to acquire are claimed to be the following [2]:

1. There appears to be an insurmountable barrier in the language used

2. Japanese information services are not well exposed to European and North American users

3. Terminals and/or terminal emulators which can handle the Japanese language are not widely available outside Japan

4. Japanese domestic online services do not operate 24 hours a day.

On the other hand, the flow of information from the US and Europe into Japan has been very smooth in all facets of information dissemination. In rough terms, the use of chemical information in Japan traditionally has accounted for about 15 per cent of world usage, either in terms of hard copies or in terms of online services. The difference is remarkable, and the reasons for the easy flow into Japan are exactly the opposite of those listed above, i.e.

1. The translation of European languages into Japanese is not a problem; many users of chemical information are bilingual with regard to technical terms

2. There are well-established distribution channels in Japan for the American and European information sources and they are operated by local organisations, book dealers and information agents.

3. There is no terminal problem. When needed, Japanese-language emulators have been developed, such as GPOT which is a counterpart to STN Express.

4. All the major online services are available in Japan 24 hours a day.

Many of these factors are related to marketing activities, but marketing is not all that matters. In what follows, Japanese activities for database production, information dissemination and language handling will be reviewed. Some difficulties and problems will be indicated although they are not unique to Japanese databases.

II. Database production in Japan: bibliographic databases

The first secondary information service was the printed abstract journal, 'Complete Chemical Abstracts of Japan', which began in 1927, abstracting back to 1877. Because it covered only Japanese literature and was printed in Japanese, the journal was not exported. A more comprehensive abstracting service was Kagaku Gijutsu Bunken Sokuho (Alerting service of Scientific and Technical Information) published first in 1958 by JICST [3] (Japan Information Center of Science and Technology) also in Japanese although it covers world literature. It is now available as the JICST file on the JOIS Online Service and its English language counterpart covering only Japanese literature is available as JICST-E on STN International. JICST is one of the three public service centres of STN.

JICST produces abstract databases of scientific meetings in collaboration with academic societies both in Japanese and English.

There are other bibliographic databases in which chemical sciences are involved, as shown in Table 1. The ^{13}C NMR data and other physical properties of synthetic polymers make an English-language database distributed in the medium of CD-ROM by Prof. Fujiwara of the University of Tsukuba. A spectral database of small molecules (IR, Massm and NMR) is also produced in the CD-ROM medium by NMCR [4] and is made available online within the laboratories of MITI. QCLDB [5] (Quantum Chemistry Literature Database) compiled by the Institute of Molecular Science contains citations to world literature on quantum-chemical *ab initio* calculation with keywords on the method of calculation, the molecular properties calculated and the kinds of basic functions used in the calculation. QCLDB has been well received by researchers at European and American universities. CHEM-J [6], which is made available by JAICI [7], is a database of titles in English in the Japanese chemical literature, and forms a catalogue of Japanese journal articles for ordering photocopies and translation.

NACSIS [8] produces several bibliographic databases: an Index of PhD Dissertations at Japanese universities, Ongoing research projects run on grants from Mombusho [9], Japanese non-Governmental Foundations and other grant-in aid in seven other countries, Abstracts of scientific meetings and Full text of academic journals on chemical sciences and electronics. The largest database of NACSIS is the Library Holdings database.

There are small-scale bibliographic databases produced at the laboratories of national universities which are made available online via an academic network. These databases are based on private collections of literature compiled by researchers in the course of their research activities.

III. Database production in Japan: factual databases

Table 2 lists factual databases produced in Japan that contain chemical information. Here again most databases are being produced at governmental institutes and national universities. There are different kinds of factual databases: numeric, graphic and textual. Other than JICST's databases which are loaded on the JOIS-F [10] online service, all the numeric and graphic databases are in the English language with only a few exceptions; they use a universal language — the numbers and the units. As is apparent from Table 2, factual databases are of a small scale, typically several thousands of records. However, they are necessarily 'knowledge-intensive', i.e. the amount of information contained per byte is far greater than in an average bibliographic database. It also means that the co-operation of active researchers is critically important for producing a useful database. What usually happens in the beginning of a numeric database is that a small group of researchers gradually builds up a collection of data in the field of their interest in card form or in other forms. Many good data collections are private until someone comes across such a collection and persuades them to make it publicly available in a printed form such as the Landolt-Boerstein Tables or in a computer-readable form including online retrieval. Researchers are not experts in information processing and they have little knowledge as to how the collection of data can be made into a database. Therefore, an approach by an information scientist and engineer is needed to have a database started.

NRIM, the National Research Institute of Metallic Materials, produces a database of mechanical properties (tensile strength) of metals and alloys, which is a component file of JOIS-F [10]. This database has a good reputation with regard to the reliability of its data. However, as in similar databases, there is a difficult technical problem, which has not been solved, of identifying materials. In the case of alloys, for example, chemical composition is not a unique key because thermal and mechanical pre-treatments usually produce very different properties. Carbon steel is a well-known case. Therefore, one can never be certain whether two alloys of apparently similar chemical composition reported in a separate paper are the one and the same material. Plastics is another type of material which needs special attention. A practical solution to this difficulty would be to use the chemical composition as the primary key and the method of producing the material as an auxiliary key. Such consideration is important in using more than one database in combination for searching properties of specific materials. Standardisation of characterisation of materials is very urgent.

Duplication of effort in the production of databases has been a matter of concern and discussion for some time. It is especially a problem in the case of factual databases because the information density is large and researchers' time and labour are involved. Thus, there are at least four major databases on infrared spectra, four mass-spectra databases and three ^{13}C NMR databases in the world. While there are small differences in these databases in terms of coverage, quality of data, file structure, etc., the nature of the work involved in building the databases is about the same. It would be more economical and useful to users if the separate efforts were integrated to reduce duplication of entries and to make the combined database larger.

CDS, Chemical Data Service, a subsidiary of a newspaper company, Chemical Daily News Service, is offering a Japanese-language database of chemical substances which are manufactured in Japan. It contains such information as chemical structure and other identification data (CAS Registry Numbers, molecular formulae, chemical names, trade names), physical properties, manufacturers, packaging, and flags to regulatory legislation.

IV. Japanese patent information

As much as 58% of CA abstracts of patent document came from Japanese patents in 1992. Whereas this does not necessarily mean that Japan issues so great a proportion of world patents, the importance of Japanese patent information has been increasing in recent years. JAPIO [11] (Japan Patent Information Organisation) is offering its own online service under the name of PATOLIS on which a Japanese patent abstract file and the INPADOC file are loaded. JAPIO also produces English language abstracts, PAJ (Patent Abstracts of Japan), both in printed form and in computer-readable files. The latter has been made available by Dialog as the JAPIO file and will also be made available by STN International. JAPIO stores all the patent documents in page image form on optical discs and offers a document delivery service based on them.

In an attempt to cope with an avalanche of patent application, the Japanese Patent Office has been implementing an in-house 'paper-less' project for some years and two recent outcomes in that direction were: (1) the installation of electronic filing of applications, which began in 1991 and (2) the decision to convert printed Kokai patent documents to CD-ROM. Since January 1993, JPO has been issuing Kokai documents on CD-ROM and will cease publishing the hard copy form at the end of 1993. JPO says that no back issues of the CD-ROM version will be available.

V. Information of regulatory substances

There are laws and regulations in developed countries for the control of toxic and other substances. It is called the 'Kashin Act' in Japan with a list of 'Existing Chemical Substances' and its additions, which are collectively referred to as ENCS [12]. ENCS contains about 32,000 names altogether. Probably because this law was enacted earlier (in April, 1974) than the analogous laws in other countries, the ENCS Inventory contains many generic and incompletely-defined names such as Formic acid salt, and these names make the list difficult to search for a specific substance. Officially these substances are described by their Japanese names only.

JETOC [13] produced a computer file of ENCS by adding molecular formulae, CAS Registry Numbers, English-language names and toxicity data (concentration and bio-degradability for some 800 substances). The file is made available online by JETOC to its members and 'search on request' orders from non-Members are also accepted.

CDS Online has flags to regulatory lists, TSCA, EINECS and ENCS. In addition, the Chemical Daily News Service recently published English-language indexes to the ENCS list in which generic names have been expanded to specific names to a practical extent. By 'practical' is meant that when a generic name corresponds theoretically to an infinite number of specific substances, the expansion is limited only to those substances which are considered to have commercial interest. The indexes are available also in magnetic tape form.

There is a plan to add a flag to the ENCS list to the CAS Registry File and CHEMLIST File on STN International.

VI. Full-text databases

The Chemical Society of Japan (CSJ) is now producing a full-text database of one of its English-language journals, the *Bulletin of the Chemical Society of Japan* (BCSJ), in the SGML format [14]. CSJ began the project of converting the production system of BCSJ from conventional CTS (computer typesetting) to database publishing in 1990 and developed DTD [14] (Data Type Definition) for BCSJ in co-operation with the American Chemical Society. The DTD is intended to be applicable to any chemistry journals with essentially no changes and to journals in other disciplines with minor modifications. The hard copy journal of

BCSJ has now been produced using TEX as the page composition system based on the SGML file since January 1993. Figures, mathematical and chemical equations, structure diagrams and tables are stored as images in separate files. The BCSJ file will be loaded on STN International as a component file of the CJO cluster.

The SGML file and the image file were integrated to make a small system in the CD-ROM medium which can be used on a workstation or on a personal computer as a hypertext-like SuperBook. The CD-ROM version will be made available from CSJ.

To reduce the cost of producing the SGML file, CSJ is now trying to develop a system which allows acceptance of author manuscripts on diskettes. The system, if developed successfully, will generate SGML tags on the author's file as the manuscripts are prepared, the tags being invisible to the author. Use of such a system will help develop similar full-text databases. R&D efforts are being made to integrate the graphics portion into SGML definitions at AIP (American Institute of Physics), ACS and CSJ.

VII. Dissemination of Japanese chemical information to the rest of the world

International online hosts offer various databases produced by Japanese organisations; these are well known and will not be discussed here. Comprehensive bibliographic databases such as CA, Inspec, Medline, and Embase contain Japanese information. Japanese organisations work to provide data into some of these databases. For example, Japanese patent documents and journal articles printed in Japanese are analysed at JAICI to produce English-language abstracts and index entries for the CAS database, supplementing the Japanese language activities at CAS. INPADOC obtains Japanese input from JAPIO, Derwent Publications obtains data from Nihon Gijutsu Boeki, Co. and GENBANK from the National Genetics Research Institute. JAERI (Japan Atomic Energy Research Institute) contributes its nuclear data to INIS. Similar co-operation arrangements exist in the area of factual databases, examples being electrochemical data, high-pressure data, crystal structure, etc.

JICST is operating three online systems; JOIS, JOIS-F and STN International. The first two are Japanese-language systems and are intended originally to provide world information to the Japanese community. However, these can now be accessed from outside Japan if a terminal with Japanese capability is available. JOIS is a search system for bibliographic files which are listed in Table 1 and JOIS-F is a search system for factual databases listed in Table 2. JOIS-F has a substance file, similar to the CAS Registry File in its functionality, which stores chemical substances in Japanese and English names, molecular formulae, stereo-chemical connection tables and CAS Registry Numbers. This substance file may be used as an entry point to the other data files of the JOIS-F system.

NACSIS operates an academic network which connects computers at major national universities and libraries. Experiments were done to connect NACSIS with the NSF (National Science Foundation) and the NACSIS files were made available to US users. Plans are under way to make NACSIS accessible through international VANs. Search commands of the NACSIS system are in English, although there are Japanese language files loaded on the system.

There are a number of stand-alone databases in the form of diskettes, CD-ROMs and magnetic tapes; many of them also contain loading/searching software. Thus, the NQRS (Nuclear Quadrupole Resonance spectra) database produced by JAICI contains some 10,000 entries of numeric data of NQRS frequency reported world-wide since 1950. Production of the database is supported by an international body for the collection of measured data. The National Institute for Materials Engineering (formerly National Chemical Laboratory for Industry) produces spectral databases of mass spectra and infrared spectra in the CD-ROM medium.

When Japanese database or software is designed to be run on a personal computer or workstation, it is usual to make them machine-independent or to make them adaptable to IBM PC or Macintosh environments.

VIII. Availability of US and European information in Japan

As already mentioned in the introduction, information that originates from other countries is widely used in Japan. On the surface, there do not seem to be inherent difficulties or problems for Japanese people to use such information: but this is not true. It *is* true that most Japanese chemists are bilingual; they read English- language papers without great difficulty and they write many research papers in under-standable English. However, that does not mean that they are truly bilingual. In fact, the Medline file on the JOIS online system sees a greater usage in Japan than the Medline file on STN International. This is because JOIS is offered with Japanese language commands, help messages, and user manuals. We, at JAICI, translate all the user documents including search manuals and database sheets into Japanese and provide the users with a Japanese-language help desk. Workshops are conducted in Japanese. Without such services to users, the English-language information service could not be a great success. After all, scientific sentences are much easier to read with a limited size of vocabulary apart from technical terms. Successful US and European information services use Japanese organisations as partners or agents for marketing. Language is such a high barrier in the information flow across countries. Dave Barry [15] indicates the point in an interesting way. Other than the language barrier, the use behaviour of the Japanese community is no different from that in other countries, although there is a time lag in the sense that what is happening in the US will happen in Japan after several years.

IX. Development of machine-assisted translation

There are more than ten commercial-grade translation systems between English and Japanese. They are said to have entered the third generation since 1990. Many work on a workstation with general purpose dictionaries being used to translate cata-logues, user manuals of machines, etc. as a man–machine system. That is, the available systems still require some sort of pre-editing and post-editing. Pre-editing is a procedure by which the sentence to be translated is supplemented, or at least changed, so that a computer can analyse it. Post-editing is a procedure to modify the translated sentence into a readable statement.

The commercial systems are, however, not adequate for translating chemical statements. We have been working in order to bring a selected system into practical use for chemical applications, and what we have done and are doing is to build a comprehensive dictionary of chemical terms. We now have compiled a dictionary of about 25,000 terms. Without an adequate dictionary a phrase "free radical reaction" would be translated to something like "liberal and revolutionary reaction".

The dictionary for a machine-assisted translation is very different from an ordinary language dictionary. It is not a simple collection of words but semantics is associated with it, i.e. a word is given an attribute which tells the computer the right translated word depending on how the word in question is used in a context.

Our system has come to a stage where we can use it for an experimental service. We are preparing Japanese language titles from original CA abstracts of a restricted subject area of high-performance liquid chromatography. The R&D is still in progress and once the problem of speed has been solved, an online application would no longer be a dream.

X. Conclusion

Transborder flow of chemical information between Japan and other countries has improved significantly during the past decade. The Japanese domestic VAN, NIFTY Serve, which is a Japanese-language VAN, is now connected to Compu-Serve. The academic network is connected to BITNET. Internet connection will also become available shortly. The geographical isolation of Japan is no longer a problem for information flow as it was prior to 1980.

Japan has become an important segment in the world chemical information community in both database building and usage. There is an increasing quantity of software which is adapted for NEC personal computers, the largest market share holder in Japanese laboratories.

Despite such international involvement, language will remain the major hurdle for non-Japanese as well as for Japanese people. The advance of machine-translation may solve part of the problem, but differences in culture and emotional aspects cannot be solved by technology. The marketing of information depends to a great extent on how well one understands the customer, and the difference in culture has been a factor which makes such understanding difficult. However, I am optimistic in this regard because a broader communication channel will necessarily force people to understand each other.

Database	Number of records	Produced by	Remarks
JICST	8 m	JICST	Science and technology, worldwide
JQUICK	2.1 m	JICST	Updates of JICST file including conference abstracts of academic societies
JMEDICINE	2 m	IMIC	Bibliography of medical sciences in Japan
JCLEARING	37,000	JICST	On-going research projects at 650 Federal Research Institutes
JCATALOG	0.5 m	JICST	JICST library holdings
NK-MEDIA	230,000	NIKKEI	Newspaper abstracts (Nikkan Kogyo Shimbun) for new technologies and new products
INIS	1.54 m	IAEA	Bibliography of nuclear energy (Japanese data added)
FRM	51,000	Hokkaido Univ.	Literature on dielectrics
QCLDB	16,561	Inst.Mol.Sci.	Literature of *ab initio* calculations
KIND	16,961	Tohoku Univ	Literature of metals

CHEM-J	149,000	JAICI	Chemical literature of Japan
KAKEN	57,300	NACSIS	Reports of subsidised research
GAKUI	52,500	NACSIS	Index to PhD theses of Japan
GAKKAI	70,000	NACSIS	Abstracts of meetings in Japan

Table 1: Bibliographic databases of chemical information

Database	Number of records	Produced by	Remarks
Japanese-language files			
Substance Dictionary (DC)	400,000	JICST	JICST Registry of chemical substances
Mass Spectra (MS)	65,000	JICST	Mass spectral data (m/e, relative intensities of organic compounds), data from Japan Mass Spec. Soc. merged into NIST file
Thermophysical	120,000	Kobe Univ.	Melting and boiling points, density, etc. Properties (TH) of inorganic and organic substances and mixtures
Crystal Structure	3,500	Crystal Soc. Japan	Organic and inorganic crystals (Crystallographic Society of Japan)
Regulatory Laws		JETOC	18 laws, 83 regulations, 1,787 substances
KASHIN / JETOC	50,000	JETOC	'Existing and New Chemical Substances', KASHIN act inventory
Metals (ME)	35,000	NRIM	Strength data of structural metallic materials
CD-NET		CDS	Chemical products, physical properties and manufacturers.
English-language files			
QCBDB	1,944	QCDB Group	Basic functions in quantum chemistry calculations
SEDATA	4,087	Tohoku Univ	Solvent extraction data
CHMGRM		Tokyo Univ	Structure of organic compounds and gas phase radical reactions
EROICA		Tokyo Univ	Physical properties of organic compounds
CNMRP	2,349	Kyoto Univ	C-13 NMR of polymers
ECDB	1,300	Yokohama Nat. Univ.	Electrolytes and electrode reactions
NQRS	10,200	JAICI	Frequencies and literature of nuclear quadrupole resonance
PAPER	5,800	Academic Societies	Full-text of journals

Table 2: Chemical factual databases

Notes

[1] **(a)** US Congressional Record — House, June 23, 1986, page H 4036 (99th Congress). 'Japanese Technical Literature Act of 1986 (Public Law 99-382); US Congressional Record — Senate, August 1, 1986, page S 10209. **(b)** The first International Conference on Japanese Information in Science, Technology and Commerce was held at University of Warwick on 1-4 September, 1987. The Conferences have been held every other year.

[2] Database Promotion Center, 'Report of Committee on Problems concerning internationalisation of Databases', 1990.

[3] JICST is a semi-governmental organisation founded by law in 1957 and supervised by the Science and Technology Agency.

[4] NMCR, the National Institute of Materials and Chemical Research, formerly National Chemical Laboratory for Industry, is one of research laboratories which belong to MITI.

[5] QCLDB contains 20,600 records reported worldwide since 1978 and is distributed in the form of a magnetic tape with loading and search software which can be modified to adapt to a particular hardware/operating system. Updated annually. Available from JAICI [7]. A complete list of entries is published as a Supplement to the Journal of Molecular Structure every year.

[6] CHEM-J is a biweekly alerting database on diskettes with loading and search software, covering about 100 Japanese journals for original papers, review articles, technical notes and new books in the field of chemistry, material science and biochemistry. Back issues on magnetic tapes. Available from Pool, Heller & Milne, Inc. (North America), FIZ Chemie (Europe) or JAICI [7].

[7] JAICI (Japan Association for International Chemical Information) is a not-for-profit chemical information centre founded in 1971 by Japanese chemical societies. It provides CAS with abstracts and index entries from Japanese patents and journal articles, and markets all CAS services and other databases in Japan.

[8] National Center for Science Information System is operated by Mombusho [9] with a primary objective to serve the Japanese university community. The databases it offers are included in Table 1.

[9] Ministry of Education, Science and Culture of the Japanese Government operates about 100 national universities and supervises other universities.

[10] JOIS-F is the Japanese-language online service of factual database system operated by JICST. The system is similar to CIS (Chemical Information System) in structure, having a substance dictionary (about 350,000 substances), with stereochemical connection tables for a part of the entries, and several factual databases each linked to the dictionary.

[11] JAPIO, formerly JAPATIC, is a not-for-profit patent information centre which works for the Japanese Patent Office and offers an online service called PATOLIS (Japanese language system). Its English language counterpart is PAJ.

[12] ENCS, Existing and New Chemical Substances, corresponds to the TSCA Inventory of the US, maintained by MITI, the Ministry of International Trade and Industry, of the Government.

[13] JETOC, Japan Chemical Industry Ecology-Toxicology and Information Center, is a not-for-profit information centre for the subject area supervised by MITI.

[14] SGML, Standard Generalised Mark-up Language, is an ISO Standard ISO8879 specifying a general-purpose grammar language for making a database. DTD plays the central role in SGML, which defines data elements and their attributes of the database.

[15] Dave Barry, 'Dave Barry does Japan', Random House, N.Y., 1992.

Gazing into the future of chemical information activities

Stephen R. Heller

USDA, ARS, Beltsville, MD 20705-2350 USA,
SRHELLER@ASRR.ARSUSDA.GOV

Introduction

This presentation [1] is designed to stimulate discussion of new computer-based technology which is now available and which will become available to chemists, applied to chemistry, and most importantly, used by chemists in their everyday activities from today to well into the beginning of the next century. Starting as a presentation at a EUSIDIC meeting in Heidelberg in 1988, this paper has evolved over the past five years, and no doubt will continue to evolve as new phenomena stimulate changes in the habits and activities of chemists.

As computer technology has developed, the use of computers in chemistry has expanded from simple arithmetic calculations to very broad areas of chemistry. This paper delves into some of these areas and tries to summarise the current state of the use of computers in chemistry and what the author believes the use of computers in the field of chemistry will be a little over a decade from now, which is roughly, from the years 2005–2010.

Background

Computers, like any other technological tool, have become integrated gradually into the daily routine of chemists. The widespread use of computers in chemistry has clearly been handicapped by a number of factors, a major one being the lack of familiarity with this new technology on the part of chemists and managers in the field of chemistry. This is true from academia to government to industry. In working to locate supporting facts for this article I heard this belief mentioned a number of times. Phrases such as you need to "raise a generation of people who are comfortable with these tools"[2], and "raise a generation of advocates"[3] came from professionals in the field of market research. Thus I concluded that wide-scale and heavy use of computers by chemists has not yet started.

At present, the routine use of computers in support of research and production in a chemistry lab or office, other than for word processing, spreadsheets, and literature searching, is low [4] (defined as less than 25% of the potential users). Why is this the case? There are a few hundred thousand chemists in the USA and many of them have computers. It is generally thought that most (> 90%) of these individuals have computers, virtually of all which are IBM PCs and clones or Apple Macintoshes, the remainder having Sun, Silicon Graphics, DEC or other manufacturers of workstations. It has not possible to obtain any definitive information on the number of scientists or chemists with PCs. Marketing surveys have not addressed such questions [5].

If one combines these numbers with those in other developed countries one could estimate there are today about 800,000 chemists [5] as a potential market for various computers and computer systems and for software specifically designed to support the needs of the chemist. What I hope this paper will address is why, if there are more chemists today compared to 30 to 40 years ago, is the use of chemical information by end users (fewer books bought, fewer CA subscriptions, etc.) less. Logic would dictate that with more chemists and more information there should be more access and use. Since this is not the case, what is likely to help bring the usage back to a logical or reasonable level, based on the size of the audience? This paper will examine some possible reasons why large numbers of chemists have not yet decided that computers are a necessary tool for the conduct of their everyday research and administrative work, thus explaining the lack of extensive use of computers and related computer technology.

Please note that when the qualitative phrases 'few' or 'low' (as defined in reference 4) are mentioned for the overall use of a particular piece of computer, a computer program, or a computerised database, the phrase is meant in comparison to the overall potential purchase and use by some 800,000 potential users (chemists) worldwide. With the exception of one series of marketing studies in the area of computational chemistry [3], there have, to date, been no published studies of the use of computers and related computer systems by the end users in the chemical community. (Studies on the use of computers and databases in libraries are not regarded here as end-user studies.) The reason for this is the small size of the current market which does not justify the investment for such a survey [2,3]. Thus the reader will have to accept the lack of hard statistics for many of the statements presented here.

While selling a total (over the lifetime of the program) of a few hundred or even a few thousand molecular modelling or structure drawing programs is, today, a major accomplishment in the business of software for chemistry, it is a minor event relative to the daily sales of word processing, database management, spreadsheets, and other such programs. The lack of any public software companies in chemistry (i.e., software companies devoted exclusively to selling software for chemistry and whose stock is available to the public) is indirect evidence to support this position that there is, at present, no major financial incentive to go into this business.

Before proceeding to the main thrust of the speculations into the future of computers in chemistry it is important to note that there are some labs as well as areas of chemistry in which the use of computers is very high. As mentioned above, the area of computational chemistry is clearly one of these areas. While it is estimated there are 1000 sites worldwide with some 2000 academic and industrial chemists now involved in this area [3, page 4], this is less than 1% of the chemists in the world. In almost all areas of spectroscopy computers are heavily used to acquire and analyse data. A reader involved in these areas of chemistry would certainly not fall within the 'low' range of computer use. However, these 'pockets' of high computer use, when averaged with the entire chemistry community, I believe are consistent with the levels of usage stated here.

Computer and chemical information issues

Table 1 summarises both the issues which are to be discussed here as well as the current and predicted level of activities in these areas. Space in these Proceedings

does not permit a full analysis of all of these topics. Thus a few representative issues
will be mentioned. Tables 2-12 list details for many of these issues.

Topic	Today	2000
Computer literacy	Low—moderate	Moderate—high
Computer chip technology	Intel 386,486; Motorola 68000; RISC	Intel 986; Motorola 98000; RISC
Operating systems	DOS, UNIX; Windows; OS/2; Macintosh	Mostly enhanced, friendly, UNIX
Telecommunications	Moderate usage 2400-9600 baud speeds	Heavy usage 1 million++ baud speeds
Interfaces	Offensive/exacting	Transparent/voice based
Graphics	Low usage in most software	Predominant usage in most software
CD-ROM	Low end-user usage	High end-user usage
Chemical information	Raw & unprocessed	Processed & analysed
Online usage for chemistry	Low	Low
SDI	Manual or by post	Electronic
Databases	Bibliographic	Numeric & factual
Beilstein	E-V Series being published	E-V Series still being published
Chemical catalogues	Online searching of catalogues	Online ordering from catalogues
Chemical identification	CAS RN & BRN	Chemical structure
Molecular modelling	Few	Some
Educational software	Random; not integrated with textbooks	Integrated with textbooks
Publishing	Semi-Electronic	Mostly Electronic
Books	Thought of as probable dinosaurs	Thought of as probable dinosaurs
Instruments	Semi-automated	Fully-automated with ISO data transfer standards

Table 1 Issues for discussion

Today	Networks being used routinely by many chemists. BITNET, CSNET, EARNET, Internet, JANET, NORDUNET, SPAN, and other networks used by scientists a few times per week. Some companies have internal networks for many of their end-user PCs. Telecommunications speeds in the range of 2400–9600 baud.
2000	Networks and e-mail used all of the time. Automatic interfacing between all networks routine. Automatic logins for mail done everyday before the scientist comes to work. E-mail automatically re-routed as you travel to meetings, holidays, and home.
	Local-area and wide-area networks are widely available within most organisations. Large databases more readily available within organisations. Telecommunications speeds in the range of 2.4 million + baud.

Table 2 Telecommunications/Networks

Today	Programs in their infancy.
2000	Voice control for input with lots of graphics. Standards for graphics and data are common. IUPAC, CODATA, ASTM, ISO, and other organisations agree on data transfer protocols.

Table 3 Interfaces

Today	Usage in its infancy. Lack of compatibility. Lack of standards. FAX transmission in its infancy.
2000	Graphics software packages are widespread. Graphics routinely sent electronically (Microsoft Chart) PCs have built in FAXs for receiving and transmitting chemical structures and tables of data.

Table 4 Graphics

Today	Chemistry CD-ROM products are rare today. Low density (600 MB) CDs. e.g., Aldrich MSDS, Beilstein Current Facts, Canadian Toxicity Databases, NIST Mass Spectrometry, CAS 12th Collective Index, C&H — Dictionary of Natural Products
2000	New products and high density CD-ROMs (6 Billion + Bytes) Heilbron Dictionary of Organic Chemicals CRC Handbook, CAS volume(s) on CD-ROM Subsets of CAS, Beilstein, & Gmelin Collections of numeric databases (e.g., IR, NMR, & MS databases from NIST & Chemical Concepts) Most Journals

Table 5 CD-ROM

Today	Most information is raw, unprocessed, and un-evaluated (CAS, Beilstein, VINITI) — most abstracting and data extraction is done in-house
2000	Greater reliance on processed and evaluated data, such as Beilstein, Gmelin, IUPAC data series, CRC Handbooks. CAS, Beilstein, VINITI — economic factors will cause most abstracting and data extraction to be done by free-lance workers at home. Articles and abstracts all sent electronically from abstractor to abstracting service.

Table 6 Chemical Information

Today	Popular feature for vendors. Results mailed to customers or left for online downloading. [6]
2000	Popular feature for vendors. Results automatically sent electronically to customers' PC via networks. Customers can order SDI articles of interest electronically.

Table 7 SDI

Today	Still in the age of bibliographic databases.
2000	Second generation of databases — numeric and factual data overtake bibliographic databases in usage. Usage increases as scientists realise need for (good) data for dry lab work (modelling, etc.)

Table 8 Databases

Today	Lots of printed catalogues. A few catalogues on disk or CD-ROM (e.g. Aldrich and Kodak)
2000	Catalogues on CD-ROMs. Users order directly over the phone from their labs. Ordering by credit card is routine.

Table 9 Chemical Catalogues

Today	CAS Registry Number reigns supreme.
2000	With chemical structures in all important databases, special identification numbers have little use. Standard molecular data formats allow for interfacing between all public and private files.

Table 10 Chemical Identification

Today	Random usage. Software used in teaching high school and college chemistry does not come with textbooks, but as separate products.
2000	Software integrated into textbooks [7] (G. D. Wiggins, 'Chemical Information Sources' [8]) PC floppy disk programs part of all undergraduate texts. Chemical Information courses have PC based tutorials and practical online sessions.

Table 11 Educational Software

Today	Journal articles are almost the only socially acceptable form of communication and reward/promotion. Some scientific manuscripts submitted in electronic form, but process is neither widespread nor practical. Virtually all refereeing done by postal system mail, with some done by FAX.
2000	Printed journals still predominate, but electronic data submissions, electronic journals, software programs are now part of academic, government, and industrial chemist's reward/promotion system. Leading journal publishers use electronic submissions to speed up processing of publications, easier data extraction, and overall quality improvement. Electronic (FAX and e-mail) peer review predominates.

Table 12 Publishing

The heart of the matter is computer literacy. Growing up with, being familiar with, and making regular use of computers and computer systems of information will not become the norm without the necessary atmosphere and background being part of your upbringing. As mentioned in the introduction, the initial use of computers by chemists (and other scientists) was limited to performing simple calculations. Hence it is no surprise that the area of chemistry in which computers have been used is primarily computational chemistry. But the usage even in this area is low. As Casale and Gelin [3] point out, "as a scientific discipline, computational chemistry is in its infancy".

The current state of education in many parts of the world will make further usage difficult. However I would hope that in college and graduate school there would be sufficient competence to train the upcoming generation of chemists to become very familiar with computers, through the introduction of computer application courses taught by chemists in chemistry departments. Without an increase in the level of computer literacy the remaining issues are pretty much irrelevant.

There are two facets in using computers; writing programs and using programs. The writing of programs is really a rather limited issue. A computer is a tool. When a chemist gets too involved in the tool then he or she is, more often than not, no longer doing chemistry. What matters is using programs. To do this effectively and properly you need to know what a computer can do for you in the area in which you need to solve a problem. I do not need to be an automotive engineer to know that to get somewhere by car, I need a car, and need to know how to drive it, and know where I am going. The same is true with computers. Understanding what a computer can, or cannot, do is the important step. Then either finding software and hardware to do it, or getting someone to produce what is needed to get the job done, is relatively simple. I believe that virtually no chemists use computers as an end in themselves and that chemists should use computers as one of many tools to do their job, but only if the computer is the most effective way, and not a barrier, to do the job better, more effectively, and more efficiently.

Most chemists use computers for only administrative purposes (such as writing a manuscript which may or may not include chemical diagrams). I would argue that the reason for the lack of extensive use of computers is that the majority of computers (PCs of the 8086, 8088, 286, and 386 vintage) which are readily available to the chemist are of insufficient capacity and capability to do effective work other than word processing, structure drawing, and spreadsheet calculations. (Without the available computers moreover, there has been no incentive to develop the software for chemists.) Until just very recently the computers with the necessary cpu speed (e.g., 486 cpu PCs, DEC, Sun, Silicon Graphics, and other type workstations) and available disk space to do a variety of scientific applications (modelling, quantum chemistry calculations, spectral interpretation and prediction, database searching of spectral data, image analysis, etc.) were much too expensive for most individual chemists to have on their desks or in their labs. As little as 2 years ago a computer with an Intel 286 cpu and 40 MB hard disk was considered a state-of-the-art computer system. Almost nobody with a PC would keep a mass spectral database [9] and search system, requiring some 23 MB of hard disk space on a computer system with a 40 MB hard disk. Today, to run a modern PC operating system (using DOS 5 and the Windows or OS/2 operating systems) one needs at least an Intel 486 cpu (with a 50-66 Mhz clock) and 300-500 MB of disk space. A recent article from a monthly computer magazine [10] added up the disk storage requirements for a little over a dozen pieces of popular business oriented software and the total of the disk space required came to almost 100 MB, not including any of the disk space required for program swapping or any of the space needed for files of data and information.

In the next few years chemists will be able to replace existing low power (e.g., 286 or equivalent type of PC) computers or buy new ones with the computer power of an Intel 486/586 (Pentium) cpu (or their Sun, Silicon Graphics, DEC, or equivalent) with sufficient disk storage space to readily run complex and powerful programs.

The low usage of computers by chemists in the recent few years may be attributable to the lack of affordable adequate hardware, but it will take a number of years for this new and more powerful hardware to work its way into the system and into everyday use. Furthermore, unless software prices follow those of hardware, it is difficult to believe that many chemists will pay $2000 for a computer system and then spend thousands of dollars for additional software packages. Only low cost, high volume software is likely to succeed in the future. An experiment in mass marketing to the chemistry community is now being undertaken by Autodesk, which is hoping to increase the number of scientists using PC-based molecular modelling packages from a world-wide total of 5,000 to 100,000 or more [3]. Included in Autodesk's effort is a multi-million dollar grant program for encouraging university use of its HyperChem molecular modelling software product.

Telecommunications, networks, and e-mail

Computers are used for electronic communication by a small, but growing number of chemists. Among the reasons for the low usage are the lack of modems and dedicated phone lines as well as the difficulty in finding where people are located or information is located and initiating communication. There is also the lack of computer addressees on the necessary computer networks (Internet, BITNET, MCInet, Sprintnet, CompuServe, etc.) and the problem of connecting between networks.

With computer networks, there is no readily available phone book, no operator. While work on this problem is moving forward it is still years away from having achieved a solution.

Today, practically all numbers (actually computer network addresses) are unlisted. However I can see changes coming. A few years ago a business card had a name, title, address, and phone number. Today many business cards have FAX and Internet addresses. This is part of computer literacy. This is progress. I believe it will still take years for chemists to make routine use of Internet and the related networks connected to it. From discussions with a number of people, the estimated usage of Internet by chemists was in the range of 10-15% of those who have a computer and can access computers outside their organisations [4]. Perhaps with the new political administration in Washington DC promoting e-mail addresses on Internet (e.g., Bill Clinton can be reached at PRESIDENT@WHITE-HOUSE.GOV) computer literacy will move forward a bit faster.

Electronic mail or e-mail is slowly (due to a lack of knowledge about it) becoming a new type of network for chemists, as well as other scientists [11]. It is not an 'old-boy' network or 'invisible college' because it allows anyone accessing these systems to be an 'equal' of anyone else on the system. Electronic bulletin boards, discussion groups, and news groups, dealing with all subjects, are slowly sprouting up everywhere in all areas, each with dozens to hundreds of users. Of almost 800 such news group surveyed by Kovacs, less than 100 are in the physical sciences and of these only 20 are in chemistry [12]. Again a small percentage is observed when the topic relates to chemistry. For some examples of Chemistry list-servers or news groups see Heller's recent paper [13].

In a 1990 survey [14] it was estimated that some 10% of the overall working population in the USA and Canada uses e-mail systems, vs., only 1.3% in Europe. The ACS Computers in Chemistry (COMP) Division now distributes its newsletter via e-mail on Internet, as well as hard-copy. In mid-1992 a little less than 10% of the COMP members received the newsletter electronically [15], a number comparable with the survey mentioned above. By 1994 this survey estimated the usage in North America would grow to almost 29%, while in Europe the usage would expand to just under 5%. Certainly there are cultural differences between those two areas in the use of the telephone and modems, but the European PTTs and their policies add to the difficulty of use. I would expect the e-mail usage in chemistry to be higher today, and that e-mail will become a necessity by the end of this decade. The low cost, ease of use, and ability readily to send written information to colleagues around the world make this an ideal replacement for the existing 'old-boy' network of phone calls, meetings, and letters. For anyone, from a Nobel prize winner to a undergraduate student, to be able to communicate freely and easily, and to see what topics and areas are of current interest should improve scientific communication and research work. E-mail will be able to reduce the time for papers to be sent back and forth. Once the graphics problem is solved (both the technical standard for graphics and the speed of transmission of graphics) and put into practical use, e-mail will allow for real electronic journals [16]. This area has a great and important future for chemists throughout the world. For more details about networking, e-mail and electronic publishing please refer to the paper by Garson in these proceedings [17].

Interfaces

Another major problem with computer programs is the difficulty associated with their use. Pacman and Nintendo (the popular video games of the 1980s and early 1990s) never came with manuals. Some manuals seem more designed for weight lifting than explaining how to use a particular computer program. Rarely are manuals available in computer readable form [18]. In computerised form, manuals could be capable of being searched for a word you are interested in finding. Installing and running programs is a major energy barrier for most people. My philosophy is that if I must read the manual to use the computer program, I probably am better off without it. There is no way someone can become and remain proficient with a wide variety of programs, remembering what each does and how to perform particular tasks, as well as doing their assigned job as a chemist. Few people use their VCRs to record TV shows because they cannot figure out how to do it. This even created a market for a device which automatically sets up the VCR to record based on a set of 5 digits you type into a device. The 5 digits are published in newspapers in the USA everyday next to each TV programme listing.

Table 1 speaks of today's interfaces as being frightful and difficult. If people are not comfortable with a tool they will not use it. As computers become more powerful and better software engineers graduate and get jobs, one can only hope and expect that the interfaces in the year 2000 will become transparent and even voice based [19]. One way to accomplish this is through the extended use of graphics in computers. Today the use of high resolution graphics (1024 x 1024 pixels) is low. Colour screen size is small (12–14 inches) and expensive. By the year 2000 I would expect that every computer will have a 20 inch or larger colour monitor with at least 2048 x 2048 resolution, along with a colour laser printer or plotter with the same capabilities. With the cost of hardware decreasing this equipment should be available to most scientists in the coming decade. Related to the problem of the need for high-resolution graphics is the problem of how to transmit all the information quickly enough to be of practical use. Today's modem speeds of 2400–9600 baud are much too slow for graphics to be practical. With the current trend of better networks and telecommunications, it seems reasonable to believe that the speeds of transmissions needed for chemistry graphics will be available in the next few years.

Chemical identification and standards

In the area of chemical identification it has taken some 20 years for the CAS (Chemical Abstracts Service) Registry Number (CAS RN) to be used widely and routinely in databases and in searching for chemicals. While the CAS RN now reigns supreme for chemical identification, it suffers from the lack of any inherent intellectual value; it is, like the US Social Security number, an idiot number (notwithstanding its check digit), assigned sequentially over time. A larger number just means it entered the CAS Registry system more recently. In the past few years optical scanning devices, coupled with advances in character and vector recognition have led to the development of computer programs (see, for example the work of Johnson, et al. [20]) which are able to scan articles from the scientific literature (or from internal research reports), extract chemical information, including connection tables of chemical structures and chemical reaction data (such as solvent, temperature of reaction, etc.)

In spite of the wide use of the CAS RN in chemistry, and particularly in chemical regulation by the EPA (Environmental Protection Agency) [21], chemical names are also still very widely used for administrative and regulatory purposes. In fact the recently developed AUTONOM program [22] was initially conceived for internal processing at the Beilstein Institute for their handbook and database work. AUTONOM takes most (> 75%) [23] chemical structures and creates an IUPAC approved name for that chemical structure. Its administrative value for internal and regulatory purposes is such that it is now a commercial package [24]. Thus while there will be a need for chemical names and registry numbers, the primary need will not be a scientific one.

The ease of creating large databases of chemical structures, along with the efforts under way to create standard molecular data descriptions of molecules (e.g., the SMD (Standard Molecular Data) [25] and STAR (Self-defining Text Archive and Retrieval) [26] projects) and the increased ability to send large volumes of data over networks at high speeds, make it seem reasonable to predict that the use of the CAS RN for searching for a chemical will decrease over time. One of the major drawbacks of the CAS RN (and the Beilstein Registry Number as well) is the lack of these numbers in the private, and generally, confidential files of companies. It is not possible to use an internal identification number to search public files and vice versa. Only the chemical structure itself, when used as the 'search term' will be a practical way to see if a chemical is in another database. As different organisations represent their structures slightly differently, only the advent of a standard molecular representation or an interchange program (such as the recently developed program ConSystant [27]) will allow a user to readily search for related structures in another database of chemical structures.

Raw vs. processed information

Most of the data and information in the major chemical databases of the world are raw and unprocessed. The two largest collections, those of CAS [28] and VINITI [29] are bibliographic. In these two databases, whatever the author says is accepted at face value. Since almost all of the papers abstracted are refereed, either the author's abstract is used or CAS or VINITI write an abstract, based on the information provided in the publication. Only the Beilstein and Gmelin databases perform some measure of evaluation, although most of this work is really extraction of information. For example, in the Beilstein Handbook, online database, and CD-ROM Current Facts, the data are extracted. In the past there were some additional efforts made to assure that enough information was published in the original work to guarantee the work could be reproduced, and not every chemical reaction or piece of data was used by Beilstein. The Beilstein staff have never had the financial capability to evaluate the very large volume of data they process. Such data evaluation is rare, with the most well known example being that of the US NIST/SRD (National Institute for Standards and Technology, Standard Reference Data Program). With the possible exception of the Mass Spectrometry database [9], none of the NIST databases are of any significant size in comparison to the numbers of chemicals for which there is data available. Beilstein performs a 'second' and valuable peer review, albeit too late to keep questionable or poorly defined or unexplained science from being published. In any event, today, due to the high costs of labour, both the Beilstein Institute and VINITI have fewer in-house staff than in past years and rely more on part-time outside workers. With some 65% of the costs at CAS being labour, it is reasonable to believe that CAS will be moving in this direction again. (CAS once had predominately all of the abstracting done by outside chemists.)

CD-ROM

CD-ROMs are computer hardware devices that are just beginning to find use in chemistry. Again the problem of the lack of good software, adequate computer hardware, and available databases has limited the growth and use of this medium. While most of the educational science libraries in the USA have CD-ROM drives it is estimated that currently perhaps some 1% of the computers which are in chemistry labs and libraries have a CD-ROM drive [30]. This estimate has been supported by a non-scientific, non-systematic request for information which the author sent to the approximately 400 subscribers to the Chemical Information News Group (see above) resulted in two responses, one from Exxon and one from Rutgers (Chemistry and Physics Departments) [31]. In both cases these information specialists who replied indicated they know of no end users in their organisations who had CD-ROMs on their PCs. In addition to this survey, a number of vendors of CD-ROMs were contacted. All considered the sales and types of users to be confidential information. None kept track of the type of users who were buying their products. In one case, that of the Aldrich Chemical catalogue on CD-ROM it was learned that while the sales (at $25 per CD-ROM) of their catalogue on CD-ROM were under 1000, they publish 2.7 million copies of their printed catalogue [32] and distributed free of charge. Even in an area where computers are used more routinely in chemistry, namely computational chemistry, less than 10% of the customers using the molecular modelling software program SYBYL have requested to receive their software update on the CD-ROM offered by the vendor. This could be compared to the computer science community, where more than 80% of the users of Sun workstations receive their software and documentation on CD-ROM. Thus chemists clearly have a long way to go before they become as comfortable with this medium as computer programmers and computer systems staff are. At present I would consider the state of the chemists' use of CD-ROM as in its infancy, but I strongly feel, with proper pricing, that the growth curve for CD-ROM usage is likely to be exponential in the coming years, as evidenced by the use of CD-ROM in other fields, where prices are quite low (and volume is high). With the expected price of an internal (one that fits in the same space slot as a PC floppy disk drive) CD-ROM dropping to under $200 in 1994 usage should increase. Perhaps some bright marketing person will discover that offering a 'free' CD-ROM drive with the purchase of their product will do wonders to overcome the current energy barrier to buy a CD-ROM.

CD-ROMs, which today store about 660 million characters (about 330,000 pages of text), will, by the year 2000, replace many reference books and chemical catalogues on the chemists' bench and bookshelf. A few pioneers in this area, such as the Beilstein Institute in Frankfurt, Germany are leading the way to what will clearly be the library of the future. The Beilstein Current Facts CD-ROM has about one year of extracted data from the literature (without author names, titles, or abstracts), along with a computer chemical structure search system, all neatly collected on a single CD-ROM. Someday, the weekly issue of Chemical Abstracts will come to each chemist this way. Each chemist will have the Merck Index, CRC Handbook of Chemistry and Physics, ACS Directory of Graduate Research, and a few ACS journals, all on CD-ROMs. By the year 2000 it should be possible to custom order a set of books on CD-ROM. For example, the ACS Symposium Series of several hundred books could be entered into computer readable form and then books 'printed' on a CD-ROM on demand, the same way floppy disks are copied today. Using keywords or phrases you could select a set of books you might want on your bookshelf (actually your CD-ROM jukebox device), and send the order for such a disk to be mastered and mailed to you. Certainly custom made orders would be more expensive than pre-packaged ones, but, if marketed and priced favourably, should be well within the means of most chemists. Groups of chemists, such as the polymer or materials chemists could create their own CD-ROMs based on existing volumes already printed. IUPAC could create a CD-ROM of Pure and Applied Chemistry. The list is almost endless.

All that is needed for CD-ROM to become widely used and for almost every chemist to have his or her own private library (as it was in those days of yesteryear) is reasonable pricing. $2000–$5000 or more for a CD-ROM (from Silver Platter, CRC, or Chapman & Hall) is not likely to get many customers. The ACS has just released the Directory of Graduate Research (DGR) on CD-ROM [33]. The cost is twice that of the hard copy version. The reason for doubling the cost is said to be that the product is "much more valuable". The DGR even has a limit on the number of hits you can print out, for fear their mailing list business will be adversely impacted by being able to print of 700 names and addresses from this CD-ROM. At present one can only print out 50 names and addresses at a time (and of rather poor quality at that — names are in inverted ordered with some things capitalised, etc.)

The current mentality of "this is more valuable, so let's charge more"needs to be replaced by "this is more valuable, so let's reduce the price and sell a lot more". 'High volume' seems to be a phrase which has been genetically engineered out of the minds of those selling CD-ROM products in chemistry. One might wonder what the price of a PC would be today and how many would be sold if PCs were priced like CD-ROM chemistry products.

Electronic books and journals

The last specific topic to be covered in this paper is the area of books, journals, and online chemical information. In the online area it can be seen from the current usage of scientific and technical databases, the current generation of chemists is not very familiar with computers and chemical information. The costs of searching the chemical literature (including the various charges of connect time, search hits, printouts, and so forth) are high, averaging well over $100 per hour connected to a host main-frame computer. Compared to browsing through a book, journal, or an issue of the printed Chemical Abstracts, this is expensive. Most of the information is not evaluated. The details of the chemical synthesis method or the properties of a molecule or material are either not in the abstract or need to be found by reading the journal article or book chapter. With high fixed expenses in the creation of the information, due to the fact that abstracting and indexing is and, I believe, will always be a very labour intensive effort (even with such expected developments as the potentially useful software of Johnson, et al. [20]), there are two ways to recover the costs; either charge a lot of people small sums of money or charge a few people a lot of money. The chemical information industry, for the most part (and there are a few exceptions), has decided to opt for high prices. The results are what most would expect. Few of the hundreds of thousands of chemists referred to in the beginning of this article use computerised databases. Few subscribe to weekly literature searching (Selective Dissemination of Information — SDI) of online databases. The reason is primarily economic. Schools and even many companies cannot afford to have hundreds of chemists spending such large sums of money on literature and related online searching. Hopefully some of the database and vendor companies will begin to experiment with the notion of marketing to the thousands of potential users waiting for reasonably priced products. Years ago many people had personal subscriptions to sections of Chemical Abstracts, to journals, and so on. Will the computer revolution in general and CD-ROMs in particular cause history to come full circle? I believe by the year 2000 this is a distinct possibility if there are changes in the way in which vendors market their products. While books will never disappear from the chemist's desk, I think CD-ROM will become the preferred medium of distribution and use in many areas of chemical information. These areas include reference works, collections of books and articles on a particular subject, as well as chemical catalogues of supplies, software, and software updates.

As for computer-based journals, as stated in Table 13, publishing in a printed journal is now the socially accepted means of communication and leads to rewards and promotions. While the means of communication can easily change, the social reward situation is quite different. Universities and most other organisations which have peer review, use refereed scientific journals very heavily in their evaluation criteria. While I feel my career has not suffered due to the software and databases I have written and developed, I do not think this is the usual case. Experiments in journals which have a substantial portion of their activity in non-hard copy form are now starting to appear. One such case was Tetrahedron Computer Methodology (TCM) [34]. This journal died after some four years, owing to a variety of technical and non-technical reasons. A new partly online journal, the Online Journal of Current Clinical Trials (OJCCT), has just finally left the ground and is now available [35, 36]. This journal has more institutional support than TCM, and so it may make inroads in this area. Additionally there is another new journal, Protein Science [37], which is a biochemistry journal which started publishing in January 1992. Protein Science comes with a floppy disk of graphics, which the journal calls 'kinemages'. In any event, I can see that these experiments, coupled with better delivery mechanisms (for chemistry this is primarily software for the transmission and viewing of graphics), will by the end of the decade lead to a few journals making real headway towards the chemical community having automated journals. Additional examples of electronic publishing activities (journals, electronic libraries and so forth) and publishing experiments can be found in an article by Borman [38].

Economic issues [39]

The recent (and perhaps ongoing) recession in a number of developed countries of the world has led to the re-examination of how to sell products. When people do not fly, airlines lower their fares to fill seats. When people do not buy automobiles, General Motors, Ford, and Chrysler, along with foreign car companies lower the prices to stimulate sales. When hotels have occupancy rates below 50% and need 65% occupancy to at least break-even financially, hotels offer cheap rooms. There are many more examples outside the chemical information area, but it should suffice to state that the Japanese domination of the consumer electronics industry clearly shows that lower prices lead to higher volumes and generally higher profits. Examples in chemical information, I need only to mention such publications as the 11th edition of the Merck Index [40] (priced at $30) or the CRC Handbook of

Chemistry and Physics [41] (priced at $100), now in its 73rd edition. Both of these products sell tens of thousands of copies.

In chemical information there seems to be a pervasive attitude that information is valuable and prices must be high. Information is no doubt valuable, as evidenced by state and corporate intelligence gathering. However the notion that because something is computer readable or in electronic form it *must* be priced higher than the corresponding printed product is probably not the way to increase usage. Eastman Kodak has recently released it chemical catalogue on a floppy disk (similar to the CD-ROM Aldrich chemical catalogue mentioned earlier in this article.) Some highly paid Kodak employee decided to charge $20 for a copy (vs. a free multi-hundred page printed catalogue). Given the current cost of postage it is certain that the cost of printing and mailing the catalogue is more expensive than the sending of a floppy disk. Why then charge for the disk?

A recent front page *Wall Street Journal* article on the Electronic Campus [42] talked about what the future may be like in the publishing industry and in libraries in the next decade. The article described how most textbook publishers are not moving into the electronic age very quickly, if at all. One electronic product mentioned in the article described a CD-ROM about Greek and other ancient civilisations. The CD-ROM contains 25 volumes of Greek text (with English translation), a Greek dictionary, some 6000 photos, and drawings of artefacts. All this sells for $120, much less than the price of the corresponding printed products. This sort of entrepreneurial effort is likely to result in increased sales of such materials and is likely to be what lies in the future. Clearly, as the article points out, textbooks are getting so expensive (and are usually heavier than a portable computer containing much more information) that students are buying fewer textbooks. Those books they buy are often at used book stores (where the publisher does not get a royalty) since the cost of new books is now so high.

In 1978 the total annual online information (scientific and non-scientific (primarily legal) information) revenues were about $40 million [43]. By 1990 this had grown to an annual rate of $690 million. The most successful computer chemistry software company, Molecular Design Ltd. (MDL) of San Leandro, California, in roughly the same period of time has seen revenues go from $0 to about $50 million per year. Molecular modelling companies, of which there are at least a half dozen, together probably have total annual revenues of less than current MDL sales. (Sales revenues are based on software sales and exclude hardware and consulting/consortium groups.) Compared with other industries and especially compared to other areas of the computer industry, these revenues are rather low and these are not impressive figures [44]. I would hope that companies in this field will begin to experiment with new marketing approaches which will both increase the usage of their products and reach a larger segment of the chemistry population. The Autodesk effort with its HyperChem software is one bright example in a gloomy field. Without a greater volume of usage it is possible that information will remain a commodity for only a small portion of the chemical community.

To reinforce what was said above with regard to pricing of CD-ROM products in chemistry, one of the best summaries of this matter of the economic problems and pricing for which the chemical information community has not been able to deal with was recently reported by Harry Collier who stated [45]:

"After over 20 years, it appears to us that confusion still reigns because too few people in this branch of the information business have a realistic assessment of what their market is. 'Oh, yes,' they will say on Monday, 'we have a high-value product which we sell mainly to a specialist niche marketplace.'

'But also,' they will say on Tuesday, 'we would like additionally to reach a market of thousands of eager end-users and expand our usage. And we also like to cater for impecunious academics, for under-developed nations and for small users.'"

Summary

I believe there are two main reasons for the low use of computers and computer systems by chemists — cost and ease of use. The economics of chemical information, up to this point in time, made it a tool for few users and the wealthy in the more developed nations of the world and for the more wealthy companies in those countries. More easy to use computer systems will begin to generate more usage. This will lead, slowly, to lower individual pricing schemes. This will happen in spite of the current marketing policies of most electronic chemical information providers. This should, in turn, really begin the age of lower individual computerised chemical information costs. (The classic chicken and the egg situation.) I believe that with the current and upcoming generation of hardware with the power of an Intel 486/Pentium (at 50-300 Mhz) or an equivalent UNIX-based Sun or DEC or Silicon Graphics workstation, software can be designed and implemented which will have two main features. The software will be reasonably easy to use (and be easy to remember the next day or week as to how to use a program or database system being accessed) and powerful enough to do the actual job needed to be done. By powerful I mean that the software will have the necessary 'user-friendly' interfaces (graphics, mouse, voice command, and so on), and have some AI (artificial intelligence) capability and knowledge of the subject to assist the end user in getting his or her job done. However, without close co-operation between software developers and database producers and their end users, this will not happen. Both the software and databases need to be properly designed to meet the actual end-user needs, not the needs which the vendors perceive the users have. Talking to, and more importantly, listening to, the customer or end user, is something the chemical information and related industry will have to come to grips with in the next few years if real and substantial progress is to be made for both parties.

Computers and the related technology described in this article hold the potential promise that by the 21st century more chemical information and computer systems will be available to the entire world-wide community. With larger numbers of users this should allow the costs of the products being developed to be spread across a much wider number of people, leading to higher usage, higher productivity and lower costs for all computer related products.

Acknowledgements

The author wishes to acknowledge numerous colleagues who have provided information and comments on this paper. Most of these people are specifically acknowledged in the references.

References

[1] Based on lectures given at the 10th International Conference on Computers in Chemical Research and Education (ICCCRE), Jerusalem, Israel, July 1992 and the Second International Conference on Computer Applications to Materials and Molecular Science and Engineering, Yokohama, Japan, September 1992.

[2] Tom Greeves, Daratech Inc., 140 6th Street, Cambridge, MA 02142, USA.

[3] Casale, Charles T. and Gelin, Bruce R., 'Growth and Opportunity in Computational Chemistry', 1992, The Aberdeen Group, 92 State Street, Boston, MA, USA. There is also a more extensive 1989 report published by the same organisation entitled 'Conflicting Trends in Computational Chemistry'.

[4] Ouchi, G., Brego Research, private communication. R. Venkataraghavan, Lederle Labs, private communication, T. Pierce, Rohm & Haas, private communication. J. Witiak, Rohm & Haas, private communication. H. Woodruff, Merck Labs, private communication. Low usage is defined as 0-25%, moderate usage at 26-49%, high as 50-75%, and very high as 76-100%.

[5] Attempts to find evidence or even an estimate of the number of chemists who have PCs has proven futile. The few firms that have done market surveys in this field or for computer sales in general, such as the Aberdeen group [reference 3], as well as Daratech, Inc., Dataquest, and International Data Corporation, had no information, nor did they have any idea where to get such information.

[6] In October 1992 STN began to deliver SDI results electronically, but only to an STNmail ID. *STNews*, **8**, #10, page 1, October 1992, North American Edition.

[7] In the PC computer field a regular flow of books come with computer disks. These disks are intimately related to the contents of the book. For example, a disk of DOS enhancement programs comes with the Dvorak book on DOS and PC performance. Dvorak, J. C. and Anis, N., 'Dvorak's Inside Track to DOS & PC Performance', ISBN: 0-07-881759-5, $39.95. Osborne McGraw-Hill, 1992. 2600 Tenth Street, Berkeley, CA 94710, USA.

[8] Wiggins, G. D., 'Chemical Information Sources', McGraw-Hill Series in Advanced Chemistry, ISBN: 0-07-909939-4, McGraw-Hill, New York, 1991. This book includes a 'Chemistry Reference Sources Database' of 2156 records plus the Pro-Cite search software for IBM PCs. Pro-Cite is available from Personal Bibliographic Software, PO Box 4250, Ann Arbor, MI 48106-4250, USA.

[9] PC version of the NIST/EPA/NIH Mass Spectral Database, March 1992 Version. Available from NIST/SRD, Bldg 221/A320, Gaithersburg, MD 20899, USA. The price is $1200 for the database or $200 for those who had bought previous versions.

[10] *PC World*, page 210, August 1992.

[11]. Heller, S. R., 'The Future of Chemical Information Activities', *J. Chem. Inf. Comput. Sci.*, **33**, 284-291(1993).

[12] One list of directories of academic e-mail conferences is available from the Kent State University file server. It was developed by Diane K. Kovacs and is copyright. It is available, via Internet, by ftp (file transfer protocol) from KSUVXA.KENT.EDU. The lists of directories are in the LIBRARY directory of the computer and are titled: ACADLIST.FILEx, where x is 1 to 7, depending on your area of interest (physical sciences, biological sciences, etc. File7 is the file containing the news groups in the Physical sciences and mathematics).

[13] There are new user groups being formed all the time. However it remains to be seen how long it will take for these to actually take hold and have staying power. Springer-Verlag, the company that distributes the Beilstein database, started a system for its users of the Beilstein database on the CompuServe computer system. After a year it was found that usage was too low to continue it. In the past few months user group conferences on HyperChem software, CHARMm software, Organic Chemistry (actually a restart of a system that died a few years ago), Amber software, BioSym software, and SYBYL software have been started up. IUPAC hopes to initiate one in the near future for its many members and affiliates around the world. It will be interesting to see how many of these remain and in what form they remain in 1-2 years.

[14] BIS Strategic Decisions Global Electronic Messaging Service, 1991.

[15] Pierce, T., Rohm & Haas, private communication.

[16] Meadows, A. J., and Buckle, P., 'Changing Communication Activities in the British Scientific Community', *J. Doc.*, **48**, 276-290 (1992).

[17] Garson, L., 'Data Design Issues in Creating Electronic Products for Primary Chemical Information', Proceedings of the 1993 Annecy Conference, (1993).

[18] The WordPerfect Corporation makes its WordPerfect manuals available on disk, which can be searched for words using any word processor.

[19] Calem, R. E., 'Coming Soon: The PC With Ears', *New York Times,* Business Section, page F9, August 30, 1992.

[20] Ibison, P., Johnson, A. P., Kam, F., Neville, A. G., Simpson, R., Tonnelier, R., and Venczel T., 'Automatic Extraction of Chemical Information from the Literature', page 25, Abstracts of the 10th ICCCRE, Jerusalem, Israel, July 1992.

[21] EPA Order 2880.2, 'Use of Chemical Abstracts Service Registry Data in ADP Systems', June 30, 1975.

[22] Goebels, L., Lawson, A. J., and Wisniewski, J. L. , 'AUTONOM: System for Computer Translation of Structural Diagrams into IUPAC-Compatible Names. Nomenclature of Chains and Rings', *J. Chem. Inf. Comput. Sci.*, **31**, 216-225 (1991). Wisniewski, J. L., 'AUTONOM: System for Computer Translation of Structural Diagrams into IUPAC-Compatible Names. 1. General Design', *J. Chem. Inf. Comput. Sci.*, **30**, 314-332 (1990).

[23] The figure of 75% refers to chemicals published in the current organic chemistry literature. The program does not handle stereochemistry, charged species, inorganics, peptides, or sugars. A. J. Lawson, private communication.

[24] AUTONOM, is an IBM-PC software program and costs $ 1980 (industry price) or $980 (academic price). It is available from Springer-Verlag Publishers, 175 Fifth Avenue, NY 10010, USA.

[25] Barnard, J. M., 'Draft Specification for Revised Version of the Standard Molecular Data (SMD) Format', *J. Chem. Inf. Comput. Sci.*, **30**, 81-96 (1990).

[26] Hall, S. R., 'The STAR File: A New Format for Electronic Data Transfer and Archiving', *J. Chem. Inf. Comput. Sci.*, **31**, 326-333 (1991).

[27] ConSystant is an IBM PC-based DOS program available for $199 from ExoGraphics, PO Box 655, West Milford, NJ 07480-0655, USA.

[28] The American Chemical Society established the Chemical Abstracts Service in 1907. The printed Chemical Abstracts and the related Chemical Abstracts databases are available from CAS, 2540 Olentangy River Road, PO Box 3012, Columbus, OH 43210-0012, USA. At present there are some 9.5 million abstracts in the CA computer readable database and about 11.5 million chemical structures with CAS REGN, in the structure file associated with the bibliographic database. There are some 17.5 million names associated with the 11.5 million structures.

[29] VINITI, The All Russian Institute of Scientific and Technical Information, was established in 1952. Since that time it has collected over 31 million source documents in all areas of science (not just chemistry). Of these there are some 11 million abstracts in computer readable form. Its main publication is *Referativnyi Zhurnal VINITI*. VINITI is located at 20a Uslevitcha Street, Moscow 125219, Russia. VINITI distributes its products outside Russia through Access Innovations, Inc. 4314 Mesa Grande S.E., Albuquerque, NM 87108, USA.

[30] Badger, R., Springer-Verlag, New York, private communication. Gary Wiggins, Chemistry Library, Indiana University, private communication.

[31] D. Johnson, Exxon Corporation, private communication. H. Dess, Rutgers University/Chemistry & Physics Library, private communication.

[32] Publication Department, Aldrich Chemical Company, 940 West St. Paul Avenue, Milwaukee, WI 53233, USA.

[33] 'Directory of Graduate Research', CD-ROM edition, ACS Books, ACS, 1155 16th Street, NW, Washington DC, 20036 (1993).

[34] *Tetrahedron Computer Methodology* (TCM), was published by Pergamon Press (UK) between 1988 and 1992.

[35] *The Online Journal of Current Clinical Trials* (OJCCT), a joint venture of the American Association for the Advancement of Science (AAAS) and the OCLC Online Computer Library Center, Inc. The price is $95 plus monthly

telecommunication charges. For further information contact the journal at 1333 H Street, NW, Washington, DC 20005.

[36] On 28 August 1992 it was announced (*Science*, **257**, 4 September 1992, page 1341) that OJCCT [35] has linked up with the journal *The Lancet* so that *The Lancet* could publish a printed, abridged form of a current OJCCT article.

[37] Borman. S., *C&E News*, pages 26-27, February 17, 1992.

[38] Borman, S. 'Advances in Electronic Publishing Herald Changes for Scientists', *C&E News*, pages 10-24, June 14, 1993.

[39] Heller, S. R., 'The Economic Future of Numeric Databases in Chemistry ', Proceedings of the 15th International Online Information Meeting, London, December 1991, pages 47–50. Published by Learned Information, Oxford, UK.

[40] 'The Merck Index', 11th Edition, S. Budavari, Editor, Merck & Co. Inc., Rahway, NJ 07065-0900, USA.

[41] 'CRC Handbook of Chemistry and Physics', 73rd Edition, D. Lide, Editor, CRC Press Inc., 2000 Corporate Blvd., N.W., Boca Raton, FL 33431 USA, 1992.

[42] Cox, M., 'Electronic Campus — Technology Threatens to Shatter the World of College Textbooks', *Wall St. Journal*, page 1, June 1, 1993.

[43] Williams, M., 'Highlights of the Online Database Industry', pages 1-4, Proceedings of the National Online Meeting, New York, May 1992. Published by Learned Information Inc., 143 Old Marlton Pike Medford, NJ 08055, USA.

[44] For example, the sales figure of $50 million for MDL sales after 10 years can be compared with that of Lotus Development Corporation, which was $53 million in its first year (1983). To date over 9 million copies of Lotus 1-2-3 have been sold. *Barron's Magazine*, September 14, 1992, page 12.

[45] Collier, H., *Monitor*, #148, pages 2-3, June 1993.

Data design issues in creating electronic products for primary chemical information

Lorrin R. Garson

Advanced Technology Department, Publications Division, American Chemical Society, 1155 Sixteenth Street, N.W., Washington, D.C. 20036

ABSTRACT: Until such time as digital delivery of primary chemical information is the dominant method of dissemination, electronic files will be created as by-products from the production of traditional, hardcopy publications. Publishers create journals principally based on format and appearance, whereas digital dissemination requires an underlying data structure. Online delivery of primary scientific and technical information has been mainly directed toward making full-text available without tables, graphics and mathematics. Modern broad bandwidth telecommunications, CD-ROM, and workstations with graphical user interfaces make practical the electronic delivery of more than just text. Tabular material, mathematics and special characters present a challenge as does the electronic delivery of line art, halftones and colour graphic data. SGML (Standard Generalised Markup Language) is a promising technology for providing the structure needed for electronic delivery and as a transport mechanism.

Introduction

For the past 20 years, electronic access to secondary information has been the backbone of the online industry. The availability of primary or tertiary information sources in the sciences and engineering has heretofore been limited, although there has been ever increasing discussion among scientists, publishers, librarians, and other information stakeholders for making primary information available electronically. In the past few years, electronic access to primary information has begun to appear, most notably in the availability of CD-ROM products and in the development of a variety of experimental, online systems.

The American Chemical Society (ACS) has been among the most active publishers in attempting to make its primary journal data available electronically. Beginning in 1980, in co-operation with BRS, the ACS made available online a number of issues of the *Journal of Medicinal Chemistry* on an experimental basis. This was soon followed by all the ACS journals being made publicly accessible on BRS, and in 1984 on STN International. This latter implementation is known as Chemical Journals Online (CJO) and now consists of files from several publishers: 20 ACS journals (in a file called CJACS), 7 journals from Elsevier Science Publishers B.V. (CJELSEVIER), 10 from John Wiley & Sons (CJWILEY), 18 from the Royal Society of Chemistry (CJRSC), and one each from VCH Publishers (CJVCH) and the Association of Official Analytical Chemists (CJAOAC). However, the CJO files contain only text and have met with limited success in the marketplace because of the lack of graphics, mathematics and tables, and an interface oriented toward professional online searchers rather than scientists and engineers.

With few exceptions, commercial online systems that deliver more than full-text have not yet become true products, principally because sufficient bandwidth has not been widely available for telecommunications. One exception is the joint venture between OCLC and the American Association for the Advancement of Science (AAAS) to produce the *Online Journal of Current Clinical Trials* (*OJCCT*). This publication, which has no hardcopy version, is available on a subscription basis via dial-up telecommunications with speeds up to 9600 baud and more recently via the Internet. *OJCCT* contains simple graphics (line drawings but not halftones or colour), unsophisticated tables (limited by window size) and limited mathematics. The publication has been in existence only one year, and although the number of manuscripts published has been small, it is too early to judge whether this product will be successful in the marketplace. *OJCCT* is probably the most sophisticated of todays online offerings for primary information.

There are on-going experimental projects in the scientific arena offering full-text-plus-graphics via online delivery. These are illustrated by the CORE and Red Sage projects. The CORE project (**C**hemical **O**nline **R**etrieval **E**xperiment) is a collaborative effort between the American Chemical Society, The ACS's Chemical Abstracts Service Division, Bellcore, Cornell University, and OCLC. The purpose of CORE is to create a prototype electronic library at Cornell University and to test several user interfaces and display formats. The CORE prototype will eventually contain all the ACS primary journals from 1980 to date. At present the file consists of approximately 32,000 articles from journals published in 1991-92. This prototype offers delivery of journal information as either digital images (bitmaps) of printed pages or as 'recomposed-on-the-fly' information, depending on the interface used. Retrieval of information in CORE is accomplished by creating an index of primary journal data from a hierarchical journal database and from the corresponding database records of Chemical Abstracts.

The Red Sage Project is a collaborative effort between the AT&T Bell Laboratories, Springer-Verlag New York, and the University of California at San Francisco. This project will consist of 20 journals each in the fields of molecular biology and radiology. The database is created by scanning printed journal pages to obtain bitmapped images for display and printing purposes and the bitmapped images are processed by OCR technology to create an index. CORE and Red Sage differ in that the former is based upon the pre-existence of journal data in electronic form whereas Red Sage presumes only the printed page.

Other scholarly electronic delivery projects include (1) *Post-modern Culture* in the field of contemporary literature, theory and culture; (2) the TULIP project (**T**he **U**niversity **L**icensing **P**rogram) with 42 of Elsevier's journals in materials science; (3) an on-demand printing, CD-ROM/LAN based project at Carnegie Mellon University in co-operation with University Microfilms International; (4) Primis from McGraw-Hill, which is an on-demand printing system for creating customised textbooks; (5) and the University of Virginia's Electronic Text Center project in the humanities. These projects have been described by Stu Borman in a recent article [1].

CD-ROM products have experienced an amazing growth in the marketplace in recent years. In 1986, the *CD-ROM Directory* contained just 48 CD-ROM titles; the 1992 edition contains 2012 titles and a listing of 2601 companies and organisations associated with this technology [2]. In 1988 Steve Holder [3] wrote . . .

"Less than 15 percent of the CD-ROM applications are devoted to full text replication of print media. Significantly, most of these are legal applications, typically containing the complete text of federal and state regulations. The remainder are demonstration products and various types of encyclopaedic information taking advantage of large quantities of existing text in electronic form."

There are numerous considerations for creating electronic products if they are to be successful in the marketplace. For example, electronic products should . . .

- offer some clear advantage over traditional, paper products,
- be intuitively easy to use (as simple as a pencil! [4]).
- be produced in such a manner to be affordable to users and profitable to publishers and distributors[1], and
- be based on stable technologies.

At this time it is not clear who in the future will be the producers of electronic products, but for the near term it is likely to be the same institutions which currently create traditional information products: publishers, universities, government agencies, corporations, and online vendors. Regardless of the organisation producing such products (CD-ROM, online or other), the data must be obtained in electronic form in some manner. The remainder of this paper will address this issue of data in machine-readable form and its implications for electronic product design.

Text in electronic form

To create electronic products, both search/retrieve and display modules must exist. It is possible, and perhaps even desirable, to have some navigation tool like an electronic table-of-contents so that users could electronically 'flip pages' analogous to using paper books and journals. However, this function without a search and retrieval mechanism is inadequate for multi-volume collections. Moreover, rapid search and retrieval is one of the perceived and actual advantages of electronic media. Clearly, some form of indexing is required which implies searching ASCII data[2]. Such data can be obtained either during the manufacturing process (compo-

---

[1] The issue of profitability is obvious for commercial institutions. For not-for-profit organisations, the notion of profit is not that of return on capital investment to stockholders, but rather a notion of surplus revenue to support the institution's goals. The concept of profitability and who should benefit financially from scholarly publishing is currently a controversial issue.

[2] In the future, some other character representation than ASCII my become the standard. Any new standard will have some defined bit-pattern definition to represent icons which we today call letters in English, or icons in other languages which may represent characters, objects or concepts.

(S)-(+)-Diethyl 2-Cyclohexylbutanedionate (16b). Reaction of (S)-(+)-2-cyclohexylbutanedioic acid[20b] (30 mg, 0.15 mmol) and N,N-dimethylformamide diethyl acetal (57 µL, 49 mg, 0.33 mmol) following the procedure for the preparation of (R)-(—)-**16a** afforded a clear oil (23 mg, 82%) that was not purified further. Analysis of the ^{1}H NMR spectrum indicated ~5% of the dimethyl ester was present: $[\alpha]^{21}_D = +14.9°$, $[\alpha]^{21}_{578} = +15.6°$ $[\alpha]^{21}_{546} = +18.1°$ (c 2.3, CHCl$_3$).

Figure 1a: Text without errors

(S)-(+)-Diethyl 2-Cyclohexylbutanedionate (16b). Reaction of (S)-(■)-2-cyclohexylbutanedioic acid[20b] (30 mg, 0.15 mmol) and N,N-dimethylformamide diethyl acetal (57 µL, 4■ mg, 0.33 mmol) follo■ing the procedure for the preparation of (R)-(—)-**16a** afforded a clear oil (23 mg, 82%) that was not purified further. Analysis of the ^{1}H NMR spectrum indicated ~5% of the dimethyl ester was present: $[\alpha]^{21}_D = +14.9°$, $[■]^{21}_{578} = +■5.6°$ $[\alpha]^{21}_{546} = +18.1°$ (c 2.3, CHCl$_3$).

Figure 1b: Text — 99% free of errors. Errors are indicated by the ■ character

(S)-(■)-Diethyl 2-Cyclohexylbutanedionate (16b). Reaction of (S)-(■)-2-cyclohexylbutanedioic acid■■■ (30 mg, 0.■5 mmol) and ■,■-dimethylformamide di■thyl acetal (57 ■L, 4■ mg, ■.33 mmol) follo■ing the procedure for the preparation of (R)-(■)-16a afforded a clear oil (23 mg, 82%) that was not purified further. Analysis of the ■H NMR spectrum indicated ~5% of the dimethyl ester was present: $[■]^{21}_D = +14.9■$, $[■]^{21}_{578} = +■5.6■$ $[\alpha]^{■■}_{546} = +18.1°$ (c 2.3, CHCl$_■$).

Figure 1c: Text — 95% free of errors. Errors are indicated by the ■ character

sition) or subsequently by re-keying data or by optical character recognition (OCR[1]) of the scanned, printed page.

If one is not particularly demanding about the quality of the ASCII data, OCR processing is a less expensive operation than re-keying data. However, without subsequent intervention (either algorithmic or human), only streams of text can be obtained from the OCR process, that is, 'Smith' could be an author, name of a college, or part of a reference; the term is merely a string of five characters without additional attributes. This of course restricts search/retrieval functionality. OCR scanning is also quite inaccurate, at least with today's technology. Figure 1a consists of a paragraph of text from the experimental section of an article by Stack *et al.* [5]. Figure 1b is the same paragraph containing black boxes (■) representing scanning

---

[1]   OCR is used here in a broad sense and includes functions such as topological feature
      extraction techniques, context-sensitive linguistic analysis, dynamic learning/training, etc.

errors at a 1% error rate. Likewise, Figure 1c illustrates a 5% error rate. As Figures 1a-c show, at present OCR processing will not produce high quality data. The question is, at what cost and quality will index data be collected from data on paper? Ultimately this cost must be passed on to users.

Until recently, most publishers have not retained their data in electronic form, and if they did, the data are generally in the form of 'composition tapes.' These data are in formats which are proprietary to specific computer-based composition systems. With few exceptions, such composition systems are format driven rather than being data oriented. As a minimum for search/retrieval purposes, the beginning and ending of journal articles and book chapters need to be defined and if fielded search capabilities are desired, data must be defined within specified fields. The traditional product of publishers and composition houses is the printed page and the appearance of this information on paper is a primary consideration. Fortunately, there is often sufficient formatting information in these data to enable the types of data to be identified. For example, if in a specific publication titles of papers are in a specific font, point size and centred, that information can be used to identify text associated with those formatting instructions and thus identify a particular set of characters as the title. Consistent positional information, such as authors' names immediately following titles, can also be used to identify data. But often procedural coding is unreliable because individual operators can create the same visual effect using different codes, publishers change composition houses and systems, and layout specifications are modified.

This less-than-ideal situation can be significantly improved if the procedural codes are applied consistently or even better, if pre-defined macros are used in association with types of data. *Troff* and its associated macros and pre-processors (*eqn* and *tbl*), as well as TEX and its variants, allow for such macro development. Transformations of procedurally-based data are commonly done today because this is the methodology used by most composition systems. However, it is rather costly as it requires skilled individuals to develop and maintain conversion programs, and inevitably conversions do not proceed smoothly without human intervention or even editing. From experience at the ACS, the cost for handling such data is in the range of $3.00-$5.00 per typeset page.

The potential role of SGML

If an organisation plans to use information for both electronic products and print, it quickly becomes apparent that there is a need to define data in terms of content as well as appearance. In fact, data content really needs to be separate from format but the two need to be associated in some manner. SGML[1] offers a potential for doing just this. In co-operation with the American Institute of Physics, The Royal Society of Chemistry, and the Chemical Society of Japan, the ACS is actively developing SGML technology with the intention of having ACS publications SGML-based in the future. The *Bulletin of the Chemical Society of Japan* is currently produced based on SGML.

---

[1]   SGML (Standard Generalised Markup Language) is a language which allows the content of documents to be defined and identified as unique objects.

The ACS has been preserving data from its journal production since 1975 in an 'SGML- like' format. This proprietary format is a hierarchical structure of variable length records in which data are typed as to content (titles, author, authors' affiliation, footnote, reference, table title, table caption, etc.), each sentence and paragraph are identified, and the schema allows for a broad range of non-ASCII characters. In fact, data in this format were converted to a true SGML representation for the previously mentioned CORE project. Current development in this area at the ACS principally involves designing DTDs (Document Definition Type) which define the types of data and their hierarchical relationships. SGML has the functionality to handle text, graphics, mathematics, sound, and practically any object that can be in machine readable form.[1]

Graphics in electronic form

As previously mentioned, the lack of graphics has been a serious limitation to the delivery of primary scientific information. Graphics are often considered to be all non-text information which encompasses a rather broad range of ill-defined objects. Fundamentally there are three types of graphics used in publishing chemistry journals and this probably applies to scientific publishing in general:

1. Line art such as chemical structures and schemes. In the ACS journals, line art is the preponderance of graphic material.

2. Half-tones which include items like electrophoresis and thin-layer chromatography data. In general biologically oriented or biomedical journals contain more materials of this type than other kinds of chemical journals. The ACS journal *Biochemistry* has as many half-tones as all the other journals combined.

3. Colour plates. There is an increasing use of colour in chemistry journals. In certain areas of biology and medicine, colour is much more extensively used than in chemistry. The ACS has not yet begun to examine the issue of digitised colour data for journal production although authors have furnished a number of original colour files which have resulted in excellent reproduction. In production of magazines and special publications, digital colour images are frequently produced in-house at the ACS and efforts are under way to include these digital files in the composition process.

4. Combinations of line art and grey scale. It is not uncommon to receive graphics from authors which contain photographs accompanied by text.

The traditional way to handle graphic material is to photograph art supplied by authors (or re-drawn by the publisher as required) and manually paste prints obtained from the negatives into reserved spaces on mechanicals which have been created by computer-controlled composition systems. This is the manner in which almost all large-scale publishing is done at the present. This method of manufacture only affords textural material in machine-readable form. Hence, the lack of graphics in current online delivery systems. To obtain graphic data in digital form, it is necessary to either scan the graphic images in the published journal or scan the

---

[1]   For further information on SGML the reader is referred to references [6–10]

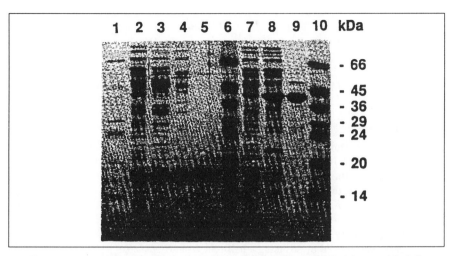

Figure 2a: grey-scale image scanned as line-art (black and white at 300 dpi).

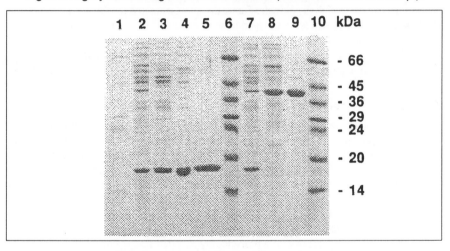

Figure 2b: grey-scale image scanned as grey- scale (256 levels of grey at 300 dpi)
followed by dithering to produce a black and white image

original art work provided by the author. Publishers rarely keep original art work
past the time a publication is printed, so scanning of graphics is generally done from
the printed journals which leads to degradation of image quality. The ACS is now
making a concerted effort to capture authors' original art work by scanning or
accepting computer generated art work directly and incorporating these data at the
time of journal production. Electronic graphic files are then included in the page
composition process. It is expected this new method of manufacturing will be
available for some ACS journals late in 1993 in time to start 1994 production and
that all of the ACS journals will use this methodology by the end of 1994. Other
large publishers are also quickly moving in this direction. The incorporation of
digital graphic data has been limited by technology, both in terms of composition

$$\int \frac{\sqrt{a^2 - x^2}}{x^3}dx = -\frac{\sqrt{a^2 - x^2}}{2x^2} + \frac{1}{2a}\log\{\{a+\frac{\sqrt{a^2-x^2}}{x}$$

Figure 3: Mathematical expression from FrameMaker samples

software being able to handle image data and affordable media to store the large graphic files, many of which can be 10-20 megabytes in size.

There is no single scanning procedure that can be applied to all three types of graphical data: line art, grey-scale and colour. Line art is appropriately represented in digital form by scanning as simply black or white.[1] Half-tones are better represented in digital form by scanning as grey-scale images with 256 levels of grey (8 bits). If half-tones are scanned as black and white, the resulting image looses most of its information content and are generally unusable. The result is similar to making a Xerographic reproduction of a photograph. Figure 2a is a grey-scale image which was created by scanning a half-tone from *Biochemistry* [12]. Even without comparison to the original, it is distorted to the point that few publishers would print such an image in their journals. Figure 2b is the same image which was created by scanning at 256 levels of grey and subsequent dithering to give a black and white file optimised for printing at 300 dpi. This latter figure is not as good quality as the original, which is to be expected, but it is a reasonable facsimile that conveys the most prominent features of the graphic printed in the journal. This treatment of the authors original art work would have been of even higher quality.

Mathematics in electronic form

The display of complex mathematics presents another challenge in creating electronic products. Most mathematical expressions are generated using complex procedural coding such as eqn in the Unix domain and TEX on a variety of platforms. In fact, TEX was originally developed by Donald Knuth at Stanford University for mathematics [13]. Figure 3 contains an example of a typical mathematical expression taken from the samples provided with FrameMaker, Macintosh Version 3.0 [14]. Such expressions could be treated like any other piece of line art and scanned from the journal to create a black and white bitmapped image. Alternatively, if the source data are preserved as a procedurally encoded file, the original file could be reprocessed to generate the image for electronic display. With either of these two alternatives one must have at hand the software which does the mathematical formula processing. Re-generating the image from the source file requires CPU cycles and potentially degrades performance, but one would obtain a higher quality image than that which could be obtained by scanning.

File size

The size of files used for electronic delivery is important. For CD-ROMs, which nominally contain 550 MB per disc, approximately 3800 pages of journal data can be stored. This is assuming full page images are scanned at 300 dpi and stored as compressed CCITT Group-4 FAX data, which requires approximately 100 KB per

[1]   For detailed information on handling graphic data in printing, see reference [11]

page, plus roughly 30 KB per page for graphic data, plus about 15 KB per page for indexes. Thus, to store the ACS's 100,000 journal pages produced each year, it would require approximately 26 CD-ROM discs, with some discs containing more than one journal title. For online delivery, file size is also important because of telecommunication considerations.

Electronic delivery of primary information can be thought of as two basic components: (1) traditional searching and retrieval, hypertext-type linking or some sort of hierarchical navigating like a table-of-contents, and (2) display of retrieved information. The display of information can be thought of in terms of two models: (1) displaying digital images of the printed page, or (2) re-composing the page from coherent, structured data much as data are processed during composition to create the original printed page. In the latter case, the goal of the re-composition process could be to reproduce exactly what the printed page looks like (not a trivial task) or create pages (or screens for online viewing) that contain all the information on the printed page. It will take time for the scientific community not to think in terms of the printed page as the paradigm for a unit of information and to gain confidence that an electronic representation can be an authoritative, archival source.

Image — Group-4 Fax TIFF Header, 300 dpi scans	1200 baud	9600 baud	56 KB	ISDN 128 KB (sec)	T1 1.544 MB (sec)
Full page, 1MB uncompressed	2 hr	14 min	2.4 min	60	5[a]
Full page 100KB compressed	11 min	1.5 min	14 sec	6	0.5

[a] *Values shown in shaded areas are deemed 'acceptable performance.'*

Table 1: Telecommunication of page images

In Table 1 are presented data for the time required to telecommunicate full page images as a function of telecommunication speeds. The shaded areas are considered acceptable performance levels. In the authors experience, users seem willing to wait about five seconds to view information on a screen before becoming impatient. The figures are given for 300 dpi images which are of adequate quality for online viewing and printing on local laser printers.[1] The images could be dynamically sampled to transmit 75, 100 or 150 dpi images for viewing in which case the transmission times would be approximately 0.06%, 11% and 25%, respectively, of the values given. However, at these lower resolutions, detail is lost and super- and subscripts often become unreadable. Thus, it is clearly unrealistic for the foreseeable future to expect that a full bitmapped journal page can be delivered in un-compressed format. As a minimum, the compressed page image would need to be delivered with local decompression. Communications via ISDN lines would give acceptable transmission and display of 100 KB compressed file images, provided decompression is accomplished quickly.

---

[1] Viewing 300 dpi images on CRTs requires horizontal and vertical pan functions as these images cannot be viewed in their entirety even on large screens with the normal 72–75 dpi resolution

Type of Data File	1200 baud	9600 baud (sec)	56 KB	ISDN 128 KB (sec)	T1 1.544 MB (sec)
Full page ASCII text only, 10KB	1 min	8[a]	1.4	0.6	0.05
Isolated table, 37KB image	4 min	30	5.3	2	0.2
Isolated table, 3.7KB ASCII	24 sec	3	0.5	0.2	0.002
Isolated math equation, 6KB image	40 sec	5	0.9	0.4	0.03
Isolated math equation, 0.6KB ASCII	4 sec	0.5	0.08	0.04	0.003
Isolated graphic, 9KB image	1 min	8	1.28	0.6	0.05

[a] *Values shown in shaded areas are deemed 'acceptable performance.'*

Table 2: Telecommunication of compound document

In Table 2 are data showing telecommunication speeds for compound documents. It is apparent that compound documents offer much improved performance for telecommunications and display, as well as requiring significantly less disk storage. The trade-off is the document must be formatted before it can be displayed. ASCII text associated with a page is about 10 KB which includes coding to define special characters and field tags to identify data elements. The underlying ASCII for tables and mathematics are approximately 10% of the size of the derived image. With a compound document and local formatting, even at 9600 baud performance would be marginally acceptable.

The future

The future for electronic delivery of scientific information looks promising from a technical viewpoint. Publishers are beginning to preserve information in electronic form and an increasing number of publishers are preserving their data in database formats, including SGML. Broad bandwidth telecommunications have become available and computer systems of sufficiently high performance are widely installed. Also, all this technology is likely to improve even more dramatically in the near future [15]. If electronic products are manufactured on sound business principles, there should soon be many successful offerings in the marketplace.

References

[1] Borman, Stu. 'Advances in Electronic Publishing Herald Changes for Scientists', *Chemical & Engineering News*, **71**, June 14, 1993, p. 10.

[2] Finlay, Matthew and Mitchell, Joanne (Editors). 'The CD-ROM Directory 1992', 7th Edition, TFPL Publishing, London, England, 1991.

[3] Holder, Steve. in 'The CD-ROM Handbook', (Chris Sherman, Editor), Intertext Publications, New York, 1988, p. 61.

[4] Petroski, Henry. 'The Pencil: A History of Design and Circumstance', Alfred A. Knopf, New York, 1992.

[5] Stack, Jeffrey G.; Curran, Dennis P.; Geib, Steven V.; Rebek, Jr., Julius; Ballester, Pablo. *J. Am. Chem. Soc.* 1992, **114**, 7007-7018.

[6] The SGML Primer, Version 2.0, SoftQuad, Inc., Toronto, Canada, 1991.

[7] Alexander, George; Walter, Mark. 'A Fresh Look at SGML: The Conventional Wisdom Changes', Seybold Report on Publishing Systems, **20**, 7, December 24, 1990, 3-16.

[8] Walter, Mark. 'Delivery Wars: Silicon Graphics and Novell Side with SGML', Seybold Report on Publishing Systems, **22**, 1, September 7, 1992, 3-8.

[9] Goldfarb, Charles F.; Rubinsky, Yuri (Editor). 'The SGML Handbook', Clarendon Press, Oxford, 1990.

[10] Association of American Publishers, Inc. 'Standard for Electronic Manuscript Preparation and Markup' (ANSI/NISO Z39.59-1988), Version 2.

[11] 'Pocket Pal: A Graphic Arts Production Handbook', 15th Edition, International Paper, Memphis, Tennessee, 1992.

[12] Meyer, E.; Leonard, N. J.; Bhat, B.; Stubbe, J.; Smith, J. M. *Biochemistry*, **31**, 21, 5022-5032.

[13] Knuth, Donald E. 'The TEXbook', Addison-Wesley, Reading, Massachusetts, 1970.

[14] Equation #217 originally from 'CRC Standard Mathematical Tables, 28th Edition', 1987.

[15] Heller, Steven R. *J. Chem. Inf. Computer. Sci.* 1993, **33**, 284-291.

Access, integration, and interoperability: the MDL approach to in-house 2D/3D searching, spreadsheets, and QSAR

Phil McHale

MDL Information Systems Inc. San Leandro, CA 94577, USA

Introduction

At the Montreux Conference of 1990, one speaker pointed out, in a discussion about the integration of chemical information systems into in-house systems, [1] that:

> "We are now entering a world of increasing heterogeneity, both in terms of hardware and software. As open system standards are more widely accepted, corporate computer environments will increasingly evolve on the 'mix and match' principle — the most appropriate tool, on the preferred hardware/software platform for any particular job, will be chosen."

This prediction has been borne out in fact, and scientists have clearly shown that they do want to 'mix and match' software packages (chemical drawing, database managers and integrators, word processors, spreadsheets, presentation packages, etc.) on a variety of platforms (Windows, Macintosh, UNIX workstations). At the same time commercial suppliers who want to address the needs of these scientists have to ensure the interoperability of their product offerings so that they can 'plug and play' with other 'best-of-breed' products on the platform(s) of choice. With this environment in mind, MDL Information Systems, Inc. (MDL) has chosen to focus its efforts on its core competency — integrated chemical information management systems — while maintaining interoperability through an open architecture, adherence to standards, and by forging strategic alliances with other hardware and software vendors in order to deliver customer-required solutions. I have summarised this approach to in-house 2D/3D searching, spreadsheets, and QSAR in the title of this paper under three main headings: access, integration and interoperability.

Access — 2D/3D searching

At the core of most in-house chemical information management systems is a chemical structure database of corporate research compounds and intermediates. This is usually in the form of 2D structures, often complemented by corresponding 3D models derived from one of the many 3D model generators. Many companies also maintain an in-house structural database of chemical reactions used to synthesise drug candidates. In addition to these in-house molecule, model, and reaction databases, there is a variety of complementary, for-sale databases covering, for example, commercially available chemicals, published and patented drugs, and chemical reactions from the literature.

Scientists ask a variety of questions relating to chemical structure, and a robust chemical search system should offer the appropriate tools to give the information needed. Typical examples are:

- Is this compound known, either internally or in the literature? [exact structure search]

- Do we have samples or is it commercially available? [exact structure search + inventory search]

- Can we synthesise it from readily available starting materials? [reaction search + inventory search]

- Do we have any compounds which contain this particular substructural fragment? [substructure search]

- Do we have any compounds which are structurally similar to the target molecule? [similarity search]

- Do we have any compounds which contain this 3D pharmacophoric moiety? [3D substructure search]

MDL's chemical searching technology embodied in ISIS provides a means to answer all these questions and several others as well. In all cases queries can be simply formulated through ISIS' intuitive forms-based graphical user interface, and can include other non-structurally based constraints. The query can be posed against one or more structure databases, which can be distributed over a network as required locally or globally.

An efficient and robust structure searching system should offer an acceptable response time, and should locate and return all the relevant hits in the database that match the query. As in-house databases increase in size, search times tend to increase, and MDL is addressing this issue in two ways. The first has been to rework our substructure searching algorithms. The new algorithms working in concert with a new index file have shown speed improvements of 7-10 depending on the query, and our current implementation in alpha test is faster than the fastest published system. [2] Our second approach to improving search speeds is to take advantage of faster processor speeds now available. ISIS/Host has been ported to run on the IBM RS/6000 under AIX, and we are also porting ISIS/Host to run on the AlphaAXP under OpenVMS. These faster processors can improve searching speeds significantly.

The issue of complete retrieval of all relevant hits was raised when 3D searching of static models in 3D structural databases first became available. The notion of *conformational flexibility* made it clear that potential hits from 3D searches might be missed if the conformer stored in the database was in a form outside the search constraints, and yet could easily adopt a conformation within the search constraints by low energy rotations around bonds. A variety of solutions to this problem was suggested, including storing multiple conformers in the database (or doing screening based on multiple conformers and storing an aggregate value), [3] building flexibility into the search query, [4] or dealing with flexibility at search time. [5] We have recently focused on the latter approach, in concert with academic and industrial collaborators, and have developed this functionality for ISIS/3D which has produced good results in evaluator-customers' hands during prototype and alpha testing.

Integration — accessing the data for spreadsheets and QSAR

Complementing and directly related to the chemical structural information discussed above will be a plethora of assay data stored in a variety of ways. Many pharmaceutical, agrochemical, and chemical companies have already constructed large relational database management systems as repositories for their assay data, and require ways of integrating this data with chemical structural information. In addition, there may be pre-existing data in legacy systems, and other information in non-rdbms locations such as system files and text database management systems. Scientists require the flexibility to be able to search, retrieve, display, and print appropriate selections and combinations of these disparate data in order to analyse the results, construct and explore hypotheses, and decide on future work.

ISIS' unique multi-database integration engine [6] enables scientists to pose questions across combined structural, bioassay, property, etc. databases through an intuitive forms-based interface. The searcher need not know where the data is physically stored (i.e., 'location transparency') nor do they need to remember the search language syntax of the underlying rdbms (i.e., 'application transparency'): ISIS handles that. ISIS integrates information from molecule, model, reaction, system file, and Oracle, Ingres, and Rdb/OpenVMS rdbms databases 'out of the box.' In addition, via a suitable interface built using the ISIS OpenGateway, data in other sources such as other rdbms (e.g. System 1032) and text database management systems (e.g. Trip from Paralog and BASISplus from IDI) can also be integrated. The technical feasibility of accessing and integrating data in online databases via an ISIS OpenGateway interface has also been demonstrated.

The retrieved results from a search are then returned to the searcher and displayed in customisable tables or forms which combine structures and data. These can be printed 'as is' and also simply transferred to word processing programs for incorporation into documents and reports (while at the same time retaining the chemical intelligence of the structures). This level of presentation may be sufficient for an initial analysis of the results, but often more manipulation is required in order to discern trends and analyse qualitative or quantitative structure-activity relationships.

The spreadsheet paradigm has become well accepted as a way to view, manipulate and analyse data presented in tabular form, and rather than spend time attempting to replicate the excellent work already done by the large spreadsheet makers (e.g. Microsoft, Lotus, Borland), MDL's approach is to present the retrieved data in a way which can be seamlessly transferred into the scientist's favourite spreadsheet, and with which he or she will already be familiar. The one aspect which current spreadsheets lack is any chemical intelligence, and we are addressing this by developing an 'R Group Calculator' to analyse the structures to be inserted into the spreadsheet. Typically a scientist will retrieve a related series of structures and their data in order to determine SAR trends. The structural data will often be presented as an invariant core structure with a series of different substituents at various attachment points. The R Group calculator will analyse the series of related structures and automatically display the core structure and substituents in a form suitable for incorporation into a spreadsheet.

Interoperability — the best tools for the job

In the discussion above, the concept of interoperability was alluded to: rather than replicate spreadsheet functionality which a scientist is already using and familiar with, we have chosen to ensure that our chemical information integration and management tools can interact transparently with any other packages which the scientist deems best for the job in hand. We have done this in a variety of ways. First, we facilitated chemical file import and export by making our file formats publicly available. [7] Second, we adhere to standards. As a simple example, our desktop products running on Windows and Macintosh platforms abide by the standards defined by Microsoft and Apple for clipboard file exchange so that ISIS/Draw and ISIS/Base can interact with other compliant programs. Data interchange and automation of procedures have been further enhanced with the introduction of Object Linking and Embedding and Dynamic Data Exchange with Windows 3.1 and Publish and Subscribe and AppleEvents with Macintosh System 7, and ISIS complies with all these standards.

A third way to ensure interoperability is to forge strategic alliances with developers and vendors of complementary solutions in order to give scientists more choice in how they solve their problems. An example of this is *Affinity MDL developers' group,* which exists to assist developers in producing products which will interact cleanly with ISIS. An illustration of this is the file conversion programs ConSystant and Chemeleon from Exographics which can read a variety of file formats into and out of ISIS.

In the area of QSAR, there are many products in use to enable scientists to analyse data and to discern structure-activity relationships from sets of structures and related activity data. Pharmacophore generators, in particular, will suggest one or more pharmacophoric groupings which seem to contribute to the observed activity, and they can be used for lead optimisation and testing the potential activity of putative bio-active compounds *in computo.* Examples include the Receptor from Tripos Associates, DrugDesigner/Receptor from Molecular Simulations Inc., Apex-3D from Biosym Technologies, and Catalyst from BioCad Corporation. [8] With all these products, exchange of structural information via MDL Molfiles is possible, and in the latter three cases the vendors have joined *Affinity* in order to develop closer integration between the chemical structure handling and data integration capabilities of ISIS and their own pharmacophore generator and other modelling and visualisation software. In a typical scenario, structural and bioassay data would be retrieved using ISIS and transferred automatically into a pharmacophore generator - any reformatting of data files or connection table formats would be done transparently to the user. The pharmacophore generator would analyse the data, with user input as required, and would produce one or more suggested pharmacophores. With a tight link to ISIS, the pharmacophore could be expressed as an ISIS structural query, and immediately posed against relevant databases (a corporate registry file and a file of commercially available chemicals, for example) to determine whether any compounds containing the putative pharmacophore were available for evaluation. [9]

Summary

We have adopted an approach to in-house 2D/3D searching, spreadsheets and QSAR which focuses on:

- chemical searching
- providing easy access to required data
- integrating information from a variety of sources and presenting the results on popular platforms, and
- ensuring seamless interoperability with other complementary packages,

so that scientists can choose the best tools to meet their requirements.

References

[1] Town, W. G. 'The integration of chemical information systems into in-house systems in a modern computer environment,' Proceedings of the Montreux 1990 Conference, 1990, pp. 69-79.

[2] Hicks, M. G.; Jochum, C. 'Substructure Search Systems. 1. Performance Comparison of the MACCS, DARC, HTSS, CAS Registry MVSSS, and S4 Substructure Search Systems,' *J. Chem. Inf. Comput. Sci.* 1990, **30**, 191-199.

[3] Christie, B. D.; Henry, D. R.; Güner, O. F.; Moock, T. E. 'MACCS-3D: A Tool for Three-Dimensional Drug Design,' in Online Information '90, 14th International Online Information Meeting Proceedings, Raitt, D. I. Ed.; 1990, Learned Information, Oxford; pp 137-161.

[4] Güner, O. F.; Henry, D. R.; Pearlman, R. S. 'Use of Flexible Queries for Searching Conformationally Flexible Molecules in Databases of Three-Dimensional Structures,' *J. Chem. Inf. Comput. Sci.* 1992, **32**, 101-109.

[5] Moock, T. E.; Henry, D. R.; Ozkabak, A.; Alamgir, M. 'Conformational Searching in ISIS/3D Databases,' *J. Chem. Inf. Comput. Sci.*, submitted for publication.

[6] patent pending

[7] Dalby, A.; Nourse, J. G.; Hounshell, W. G.; Gushurst, A. K. I.; Grier, D. L.; Leland, B. A.; Laufer, J. 'Description of Several Chemical Structure File Formats Used by Computer Programs Developed at Molecular Design Limited,' *J. Chem. Inf. Comput. Sci.* 1992, **32**, 244-255.

[8] Receptor is available from Tripos Associates, St. Louis MO; DrugDesigner/Receptor is available from Molecular Simulations, Inc., Waltham MA; Apex-3D is available from Biosym Technologies, San Diego CA; Catalyst is available from Biocad Corporation, Mountain View CA.

[9] Ricketts, D. E. '3D Database Searching Using a Pharmacophore Hypothesis as a Query: HMG CoA Reductase Inhibitors,' Scientific Applications in Rational Drug Design, Biosym Technologies Applications Note, San Diego CA.

Chem-X: the less expensive alternative

Keith Davies and Roger Upton

Chemical Design Ltd., Cromwell Park, Chipping Norton, OX7 5SR, England.

Overview

Chemical Design launched its first database product in 1988, as a project database for modelling. The pioneering 3D-database product, ChemDBS-3D, was launched in 1990 and the underlying technology was published in the same year [1]. In the spring of 1993, Chemical Design expanded its database solutions with the addition of 2D searching, reaction searching, interfaces to relational databases and client-server capabilities.

A number of strategic partnerships with database and software suppliers have also been established. As a result of this, databases of hundreds of thousands of structures and reactions are now available for searching with Chem-X. The availability of these databases and the implementation of Chem-X on PCs, Macintoshes, Unix and VMS servers offer a less expensive alternative.

Migration

Our independent market research [2] reported that many organisations were planning to review their future chemical information needs. Some companies already have well specified future hardware and end-user interface requirements. The plans for the migration from an existing information system towards one with future benefits need to consider the overall costs of staff re-training and conversion of existing applications.

One of the design goals at Chemical Design has been to use existing resources while introducing the new technologies in a paced manner. This has been approached in a variety of ways — from hardware independence through open customisation.

Platforms

Unix workstations and servers are generally considered to be the platforms on which the next generation of chemical information systems will be used. The Chem-X program operates on a wide range of platforms from desktop machines to powerful servers, and it is exactly the same Chem-X program on all. This allows a Chem-X information environment to be installed both on the computer systems primarily used today as well as those planned for the future.

Databases created and used on any of the Chem-X platforms can be moved, without change, into the preferred environment. The importance of this ability lies in being able to optimise computing resources, such as creating central databases on fast dedicated machines and then moving them to the ideal locations for general distribution. A further advantage derives from the transparent migration of databases when new hardware configurations are installed.

The supported hardware platforms are:

- PC — Windows, DOS & NT *
- Macintosh
- IBM RS/6000
- Silicon Graphics
- DEC Ultrix
- VAX/VMS
- DEC Alpha — OpenVMS, OSF* & NT*
- Sun *
- Generic X-Windows *   (* not yet released)

Customisation

The extent to which customisation of the user interface is an important issue will vary widely between different users, from those who only use simple standard screen to the more ambitious who see well designed screens are offering a significant edge. Customisation plans also need to consider many options including the role of the different hardware available, the range in degree of sophistication of the users, the degree of control over what information is to be made available, etc.

The design goals for Chem-X were:

- consistent and interchangeable layout across platforms
- screen mimic options for simulating existing capabilities
- flexible screen design for handling text, numeric, structural and spectroscopic data
- interactive editors for *ad hoc* reports
- open development tools for the rapid creation and adaptation of screens
    - layout development and design from within Chem-X session
    - full programmable script language

Desktop integration

Today there is no standard by which to transport chemical information between applications on desktop computers. Microsoft has provided standard communication tools but there is no consensus between vendors or end-users of chemical structure drawing and database software on the representations of the connectivity graph. The sending and receiving applications need to understand the dialect of the data. There is a *de facto* standard for transferring tables, for instance as used by Microsoft Excel, and graphical images. Chemical Design would welcome a non-proprietary format for chemical structure representations.

The early Windows-based chemical information software used separate applications to draw chemical structures, search databases and view the results of searches [3], with proprietary communications between the applications. Chem-X integrates these functions into one application alleviating the need for user-intervention to transfer data between the application windows. Chem-X integrates with word-processors exporting graphics as encapsulated postscript and meta-files. Links to spreadsheets, such as Excel, will be available this autumn.

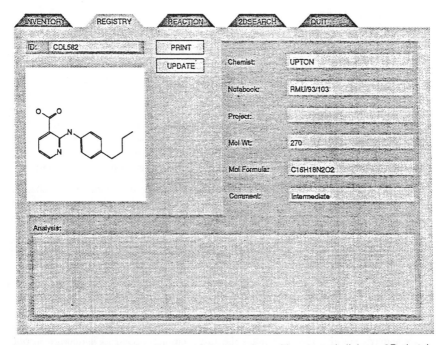

Registration screen showing the use of structure box with automatic links to 2D sketch
pad; input boxes for entering local data or remote data in a relational database; input
validation checking; automatic computation of data such as Mol. Wt. and next unique ID
number; etc.

On Unix hosts there is a standard message passing mechanism, but again no
standards exist for chemical structure messages. There are also issues relating to the
transportability of documents containing chemical structures between platforms.

Client-Server

The role of the client-server implementation is to share computer resources between
centralised servers and powerful desktop machines. The preferred choice of which
tasks are performed on the desktop client and which are performed by the server
will depend upon the computing environment itself, and so a client-server arrangement will only bring improved performance when the server CPU and disk I/O are
significantly better than a dedicated client.

Chem-X has a unique open architecture which has allowed Cambridge Scientific
Computing [4] to send queries to a Chem-X server and view the results. The simplest
mechanism is for third party clients to pass queries to a Chem-X server as an MDL
SD query file [5] and retrieve the results in an appropriate format, such as an MDL
SD file. In client-server mode the same Chem-X software runs on the desktop and
the server. This allows the user to exercise control over whether a search is
performed using the local or remote CPU. In remote mode the desktop application
can be terminated to allow the server to continue 'in batch'.

Relational database links

Chemical structure databases form only one component of the chemical information system. Relational databases often play a very central role for the storage and retrieval of data about chemical structures, including the biology, toxicology and clinical data. The integration into reports of data from relational databases with chemical structures can be problematical. It is important that a chemical structure searching and reporting system is able to exploit the power of relational systems in defining access to the information held and in optimising the search performance. When the activity tables are relational in structure, to achieve best performance, it is vital to use the relational database search engine to execute the necessary SQL JOINS rather than perform such operations in the integration software gateway.

With Chem-X, the links to relational databases make use of open communications such as SQL*NET[6]. SQL*NET provides a multi-platform industry-standard protocol for achieving the required links. A further advantage of providing such an integration at the desktop level lies in the improvements in network performance by avoiding a proprietary integration gateway which can become a bottle-neck and increase network traffic.

In specifying the links between structures held in a ChemDBS database and data held in a relational database, Chem-X makes use of standard SQL as the mechanism for defining the access to the relational data. The importance of adopting standard tools builds on the skills already available to the application developers. The SQL standards are also able to exploit the full relational power of databases such as Oracle, Ingres or 1032.

Integration with discovery tools

Tools to develop and test a hypothesis for activity are very valuable to the research scientist [7]. There may be activity and experimental data provided in central databases, while more research-oriented information on modelling or statistical analyses can be held in a project database. Chem-X databases have been designed to meet corporate and project requirements. The central database can be configured to provide both 2D and 3D search capabilities with links to relational databases while the project database many contain unvalidated data including structures which have not yet been made. The ability to access multiple databases provides the necessary integration between the central and project data.

The Chem-X worksheet provides spreadsheet-like viewing of the data and the ChemStat module of Chem-X allows data cells to be filled using standard and user-defined calculations. A range of fully validated statistical methods is provided including regression, principal components analysis (PCA), partial least squares regression (PLS), weighted least squares non-linear mapping (WLS) and clustering.

Pharmacophore Identification & 3D database searching

One commonly adopted rational approach to drug design aims to define a pharmacophore model — a geometric model of chemical properties which describes the essential shape which an active structure must be able to adopt. The 3D-database technology pioneered by Chemical Design provides a rapid tool for the automatic generation of multiple ideas for possible pharmacophores [8]. The use of 3D similarity clustering, which can group compounds into different pharmacophore

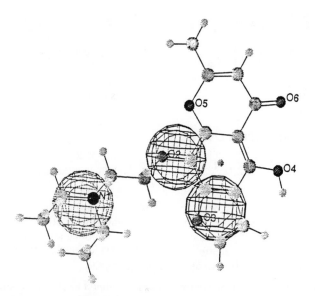

Pharmacophore model superimposed on one of the structures used in its generation

families, has been used to overcome the problems encountered when dealing with more diverse sets of active structures, that do not have a single common solution.

The pharmacophore models are used as queries for 3D-database searching. A 3D conformational search will retrieve from the 3D database all those structures which can adopt the shape as defined by the pharmacophore [1]. Significant successes have been reported using this approach to drug design [9]. In addition to the benefits to be accrued from flexible 3D searching, the transparent links to molecular modelling and statistical functionality provide further capabilities for analysing search results by similarity clustering for sorting the results and by linking into QSAR analyses.

References

1. E. K. Davies and N. W. Murrall, *J. Chem. Inf. Comput. Sci.*, 1990, **30**, 312-316.

2. Judson, consultant in computer systems for the chemical industry, private communication.

3. For example, ISIS version 1.0 from Molecular Design Limited, San Leandro, California.

4. Cambridge Scientific Computing, 875 Massachusetts Ave, Cambridge, MA 01239. USA.

5. J. G. Nourse *et al.*, *J. Chem. Inf. Comput. Sci.*, 1992, **32**, 244-255.

6. SQL*NET, ORACLE Corporation, Belmont, California.

7. For example, see literature on Renin inhibitors, by J. H. Van Drie, BioCAD, Mountain View, CA 94043, USA.

8. Pharmacophore Identification, in preparation.

9. Jon Mason, 9th Noordwijkerhout-Camerino Symposium, The Netherlands, 23-27th May 1993.

The Molecular Spreadsheet as a focal point for a drug discovery strategy

Tad Hurst and Scott DePriest

TRIPOS Associates, Inc., St. Louis, Missouri, USA

Many new tools have become available to the medicinal chemist and molecular modeller to assist in the rational design of new and effective drugs. These include a semi-automated systematic exploration of conformational space; [1] automated pharmacophore perception, [2] flexible 3D searching [3,4,5,6], 3D QSAR (CoMFA) [7] and *de novo* design techniques. [8,9]

Although these techniques each represent an important step in assisting the medicinal chemist in making better research decisions, it is often difficult to transfer information from one program to the other. It is exactly this transfer of information which is needed. For example, a pharmacophore determined by one system becomes the 3D query for a 3D searching system.

This paper discusses the Molecular Spreadsheet as a unifying paradigm for data exchange for these types of products. This allows these tools to fit together in a cohesive strategy for rational drug design. The Molecular Spreadsheet currently is used for data collection and processing for TRIPOS software including UNITY (2D and 3D flexible searching), ORACLE [10] database access, 3D QSAR (CoMFA), LEAPFROG (*de novo* design), and DISCO. These tools may be used in sequence to produce novel ideas for new synthesis or testing (Figure 1). The use of the Molecular Spreadsheet by each of the tools in this sequence is discussed in the following sections.

UNITY 2D searching and ORACLE access

The 2D (substructure) searching capabilities of UNITY allow users to find sets of structures which are related by a common fragment. This fragment usually represents the base structure from which several analogs may have been synthesised. This type of searching allows the medicinal chemist to retrieve all structures which have been made and tested in a certain area of interest. An example of the resultant spreadsheet is shown in Figure 2.

This spreadsheet also includes the biological testing data which are stored in an ORACLE database. The Molecular Spreadsheet allows easy retrieval of the testing information for each structure found in the 2D search. The rows of this spreadsheet can be graphed (Figure 3), or sorted by biological activity.

RECEPTOR and DISCO pharmacophore determination

RECEPTOR and DISCO are both programs which assist users in identifying the pharmacophore responsible for an observed biological activity. These programs

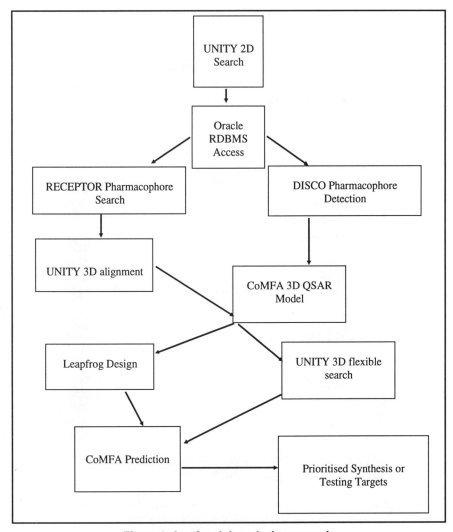

Figure 1. A rational drug design scenario

differ in the level of input required, the level of intuition required by the user, and the rigor in which the pharmacophore is determined. RECEPTOR requires (and allows) the user to enter information based on experience and intuition, and is systematic in its investigation of the conformational space of the active molecules. DISCO is a more approximate technique, and requires only the active structures as input. DISCO does not explore conformational space systematically. DISCO does align the active molecules according to the pharmacophore it determines, whereas RECEPTOR requires an additional step to align the active structures.

RECEPTOR

RECEPTOR is a systematic conformational search program developed by Richard Dammkoehler *et al.* at Washington University in St. Louis, Mo. This program

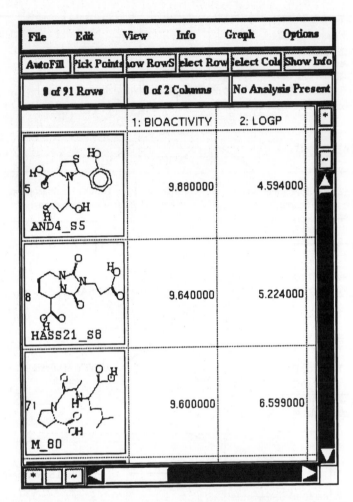

Figure 2. ACE inhibitors found using UNITY 2D searching and ORACLE data displayed in the Molecular Spreadsheet

represents a semi-automatic implementation of the active analog technique described by Marshall *et al.* [11] , and produces one or more pharmacophore definitions which are consistent with all of the active structures presented to it as its input.

Input to this program includes, in addition to the active structures, the definition of pharmacophoric points in each structure to be considered. For example, the user might have recognised that activity in a series of structures depends on the presence of a phenolic oxygen atom, an aromatic ring, and a hydrogen bond donor. The atoms which represent these functional groups would have to be specified for each of the structures.

RECEPTOR produces on output one or more sets of distances between these functional groups which are accessible in all of the active structures. Each set of

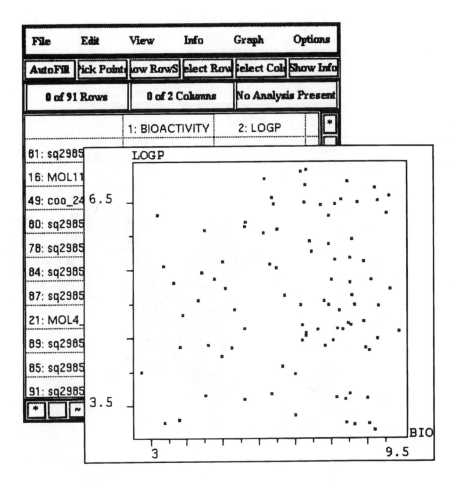

Figure 3. A graph of the ACE inhibition data from ORACLE using the Molecular Spreadsheet

distances, along with the definition of the functional groups, represent a pharma-cophore definition.

Ideally, the sets of distances produced represent a single cluster in distance space (i.e. slight variations on the same pharmacophore definition). A representative set is then chosen. When more than one cluster is produced, a representative of each must be chosen and investigated further as a possibility.

The Molecular Spreadsheet is used to display the various distances produced by RECEPTOR searching. In Figure 4, the results of studying the ACE series is shown. In this example one cluster was found, and can be used as a Query for UNITY Flexible 3D searching. The structures used to identify the pharmacophore are shown aligned according to that pharmacophore.

DISCO

DISCO is a program developed by Yvonne Martin *et al.* at Abbott Laboratories in Chicago, Illinois. This program generates pharmacophore definitions from a set of

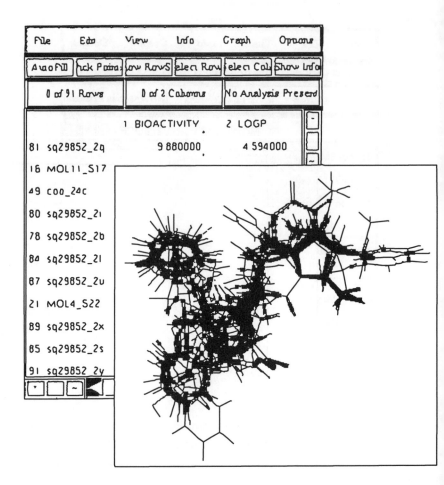

Figure 4. The results of a RECEPTOR study showing 68 overlayed active ACE inhibito

active structures, as does RECEPTOR. DISCO, however, requires very little additional input from the user, and thus is much more of an automated system.

DISCO produces for each active structure the interaction points which might be involved in binding to a receptor. Each structure may be represented by several conformations. DISCO looks for common sets of distances between systematically selected subsets of these pharmacophoric points.

DISCO produces on output one or more pharmacophore models. These are displayed in the Molecular Spreadsheet. For example, Figure 5 shows the results of a DISCO run on 28 of the active structures from the ACE series. The conformations of each of the active structures which produced each pharmacophore are aligned automatically according to the pharmacophore definition.

CoMFA 3D QSAR and UNITY 3D alignment

CoMFA is a powerful technique which produces a QSAR (Quantitative Structure-Activity Relationship) model which can be used to predict the activities of new or

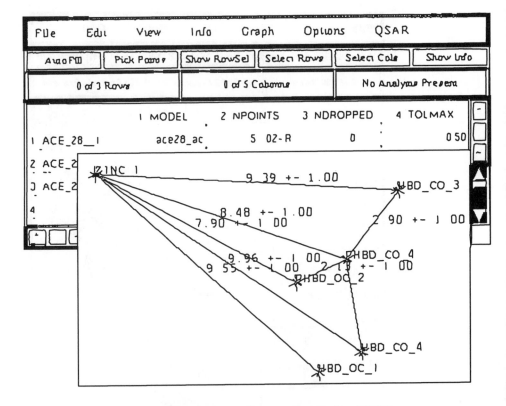

Figure 5. ACE pharmacophores suggested by DISCO

untested structures. CoMFA was invented and developed by Richard Cramer *et al.* [7] of TRIPOS Associates, Inc., in St. Louis, Missouri. This technique uses the fields which molecules present to receptors as predictors for their activity. The results of CoMFA runs can be graphed in 3D, and give a visual representation (and explanation) of the factors which control activity. Areas around the structures which should not contain any steric interaction or which benefit from steric interaction, can be visualised. Electrostatic interactions can also be displayed. The model produced can be used to predict the activities of compounds for which measured activities are not (yet) available.

CoMFA requires as input the biological activities and the chemical structures for several active compounds. The structures must be aligned according to the pharmacophore model so that the fields presented by each structure can be compared. This alignment represents the major difficulty in using CoMFA.

The UNITY product has 3D searching capabilities which can be used to align structures according to a pharmacophore definition, such as those produced by RECEPTOR. All of the structures except those which cannot attain the query geometry can be aligned for use in CoMFA. The resultant structures from DISCO are already aligned and do not require explicit alignment.

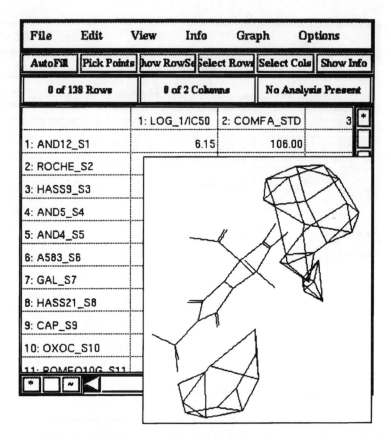

Figure 6. A graph of the regions where steric interactions reduce ACE inhibition as determined by a CoMFAt analysis

The Molecular Spreadsheet is used by CoMFA as a repository for the structures and the biological data, as well as the field information to be used as predictor variables. An example of a CoMFA study of the ACE inhibitors is shown in Figure 6. This example produces a predictive QSAR which has three components based on 120 inhibitors. This model has a cross-validated r^2 of 0.68. The steric components of the QSAR are displayed over one of the structures.

LEAPFROG

The LEAPFROG program was developed at TRIPOS Associates and is a *de novo* design technique which can generate new structures which either fit an enzyme cavity or present the field characteristics determined to be important by a CoMFA analysis. LEAPFROG uses a fragment-growth technique to continually modify and grow structures with the proper characteristics. This technique is considered an application of the Genetic Algorithm approach, and can produce structures which vary significantly from those produced by other methods.

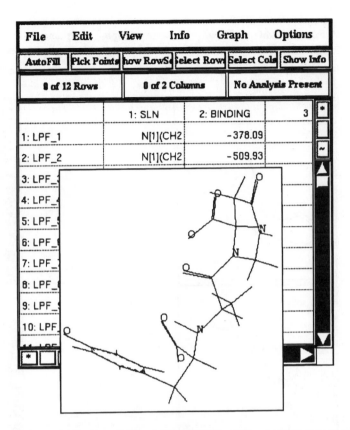

Figure 7. Structures generated by LEAPFROG to fit the ACE pharmacophore

The structures produced by this method are generally quite reasonable, because they are generated from synthetically important groups. The resultant structures can be considered candidates for synthesis.

LEAPFROG uses the Molecular Spreadsheet to track the structures as they are produced and the effective fit of the structures to the CoMFA model or receptor site. Figure 7 shows the spreadsheet with the structures generated by LEAPFROG to complement the ACE CoMFA study. One of the structures in the spreadsheet is displayed in 3D graphical form.

UNITY flexible 3D searching

UNITY allows the user to find the structures in a database which match a pharma-cophore definition. The database may be one consisting of all the structures synthesised at the user's site (the company registry), or may be one obtained from a database vendor. If the structures come from the company database, samples of the compounds may be available for immediate testing.

The structures produced by this process will include the structural analogs found by 2D searching, but also will include structures which are in completely different classes. These novel structures are still likely to be active at the same site and with

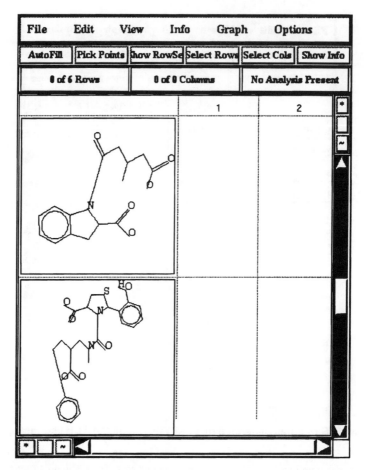

Figure 8. Possible new ACE inhibitors found by UNITY flexible 3D searching

the same mode of action as the structural analogs, and thus represent possible new lead compounds.

UNITY addresses the flexibility of the structures at search time, rather than the less rigorous method of storing a handful of representative structures. The method used is called Directed-Tweak [12], and has been shown to be the most effective and fastest of the conformational techniques available for 3D searching [13]. UNITY 3D searching also makes use of parallelisation to decrease the search times on multi-processor machines.

UNITY uses the Molecular Spreadsheet to view and analyse the structures which are found as the results of the 3D search. For example, the results of a 3D search for the pharmacophore developed by DISCO for the ACE series is shown in Figure 8. These structures represent the possible new lead compounds.

CoMFA prioritisation of the synthesis candidates

The structures produced by UNITY 3D Flexible searching or by LEAPFROG can be prioritised according to their predicted activities. The CoMFA model developed

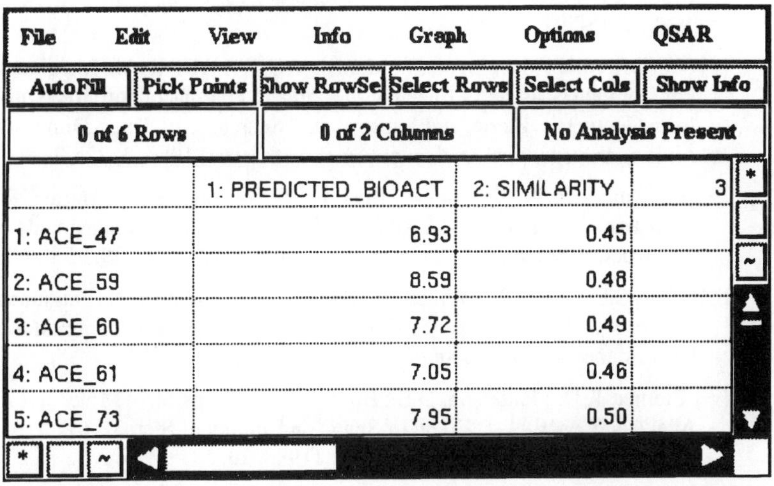

File	Edit	View	Info	Graph	Options	QSAR

AutoFill	Pick Points	Show RowSe	Select Rows	Select Cols	Show Info
0 of 6 Rows		0 of 2 Columns		No Analysis Present	

	1: PREDICTED_BIOACT	2: SIMILARITY	3
1: ACE_47	6.93	0.45	
2: ACE_59	8.59	0.48	
3: ACE_60	7.72	0.49	
4: ACE_61	7.05	0.46	
5: ACE_73	7.95	0.50	

Figure 9. Prediction of activities for the synthesis candidates using CoMFA

previously is ideal for this prediction, since both UNITY and LEAPFROG produce the structures already aligned for CoMFA analysis.

Figure 9 shows a Molecular Spreadsheet which contains the most dissimilar structures sorted by predicted activity. In this example, one column of the spreadsheet displays the similarity of the structure to the lead structure, Captopril. The structures with the least similarity are often of most interest, since they could represent novel lead compounds.

The Molecular Spreadsheet

The Molecular Spreadsheet is used by all of the techniques described above, and is the data-transfer method between the various steps. The utility of this tool extends beyond those applications listed here. It is used in many other modules in SYBYL [14]. The Molecular Spreadsheet is created in an open environment, and is thus ideal for use in other non-TRIPOS applications where columns of data related to structures must be addressed. The Molecular Spreadsheet is thus a central focus for an integration of the applications which support the drug and agrochemical discovery process.

References

[1] Dammkoehler, R.A.; Karasek, S.F.; Shands, E.F.B.; Marshall, G.R. 'Constrained search of conformational hyperspace', *J. Comput.-Aided Mol. Des.*, 1989, **3**, 3 -21.

[2] Martin, Y.C.; Bures, M.G. ; Danaher, E.A.; DeLazzer; J.; Lico, I.; Pavlik, P.A. 'A fast new approach to pharmacophore mapping and its application to dopaminergic and benzodiazepine agonists', *J. Comput.-Aided Mol.. Des.*, 1993, **7**, 83-102.

[3] Martin, Y.C.; Danaher, E.B.; May, C.S.; Weininger, D. 'MENTHOR, a Database System for the Storage and Retrieval of Three-dimensional Molecular Structures and Associated Data Searchable by Substructural,

Biological, Physical, or Geometric Properties'. *J. Comput.Aided Mol. Des.* 1988, **2**, 15-29.

[4] Van Drie, J.H.; Weininger, D.; Martin, Y.C. 'ALADDIN: an Integrated Tool For Computer-assisted Molecular Design and Pharmacophore Recognition from Geometric, Steric, and Substructural Searching of Three-Dimensional Molecular Structures'. *J. Comput.-Aided Mol. Des.* 1989, **3**, 225-251.

[5] Sheridan, R.P., Nilakantan, R.; Rusinko III, A.; Bauman, N.; Haraki, K.S.; Venkataraghavan, R. '3DSEARCH: a System for Three-dimensional Substructure Searching'. J. *Chem. Inf. Comput. Sci.*, 1989, **29**, 255-260.

[6] Jakes, S.E.; Watts, N.; Willett, P.; Bawden, D.; Fischer, J.D. 'Pharmacophore Pattern Matching in Files of 3-D Chemical Structures: Evaluation of Search Performance'. *J. Mol. Graphics,* 1987, **5**, 41-48.

[7] Cramer, R.D. ; Patterson, D.E.; Bunce, J.D. 'Comparative Molecular Field Analysis (CoMFA). 1. Effect of Shape on Binding of Steroids to CarrierProteins', *J. Amer. Chem. Soc.* 11988, **10**, 5959-5967.

[8] Chau,P.-L.; Dean,P.M.; 'Automated site-directed drug design: The generation of a basic set of fragments to be used for automated structure assembly'. *J. Comput.-Aided Mol. Des.*, 1992, **6**,385-396

[9] Böhm, H.-J. 'The computer program LUDI: A new method for the *de novo* design of enzyme inhibitors', *J. Comput.-Aided Mol. Des.*, 1992, **6**, 61-78.

[10] ORACLE is a registered trade mark of Oracle Corporation

[11] Mayer, D.; Naylor, C.B.; Motoc, G.R., Marshall, G.R., 'A unique geometry of the active site of angiotensin-converting enzyme consistent with structure-activity studies'. *J. Comput.-Aided Mol. Design*, 1987, **1**, 3-16.

[12] Hurst, T. 'Flexible 3D Searching: The Directed Tweak Technique', *J. Chem. Inf. Comput. Sci.*, Submitted Aug. 1993.

[13] Clark, D.E.; Jones, G.; Willett, P.; Kinney, P.W.; Glen, R.C. 'Pharmacophore Pattern Matching in Files of Three-Dimensional Chemical Structures: Comparison Of Conformational-Searching Algorithms For Flexible Searching'. *J. Chem. Inf. Comput. Sci.*, 1993, accepted for publication.

[14] SYBYL is a registered trademark of TRIPOS Associates, Inc.

A distributed chemical information database system

David Weininger

Daylight Chemical Information Systems, Inc., Irvine, CA, USA

Abstract

A distributed and extensible chemical information system is described. Advantages of structure-based chemical databases are elucidated which are primarily performance and database compatibility. Implementations of a thesaurus-oriented (THOR) database server and an in-memory search server (Merlin) are discussed. A number of end-user application programs are also described. Experience operating this system with large databases is summarised and it is concluded that this system provides high-performance chemical information delivery in a wide variety of environments using currently-available hardware. The system described is commercially available from Daylight CIS, Inc. [Daylight, 1993].

Introduction

Data processing capabilities are increasing at a faster rate than chemical information is accumulated. Data processing capability doubles in one to two years whereas the amount of available chemical information doubles in about 20 years. Eventually, data processing capacity must outpace our information production. It is observed that this is already the case for most practical chemical information problems. A single modern workstation has the capacity to store a corporate database; a modern multiprocessor server can store most of the world's chemical information.

As computer capabilities increase, chemical information problems change from those of getting information into and out of a machine to those of providing data which are relevant to a user's task at hand. This paper describes a chemical information system which was designed from scratch to store all the world's chemical information. Only methods and algorithms which meet that criterion are used in this system. Chemical data are keyed to structure, the universal language of chemistry. All chemical-oriented data are accessible via a thesaurus-oriented database. Database searching is done in-memory, providing a 100,000-fold speed advantage over disk-based systems. This system is now in commercial production. The implementation described here is Daylight CIS Software Version 4.32, released in July 1993.

Organisation of an extensible data system

General requirements of an information system include:

(a) describe all relevant types of information from disparate sources

(b) accurately record the identity of stored information (i.e. what specific data are known to be about)

(c) store information with extremely high reliability (archival)

(d) rapidly search for information by data content without previous knowledge of database contents (exploratory data analysis)

(e) operate in a machine-independent fashion such that the contained data are accessible to all potential users

(f) scalable with respect to size (100s to millions of entries) and user load (single user to 1000s of simultaneous users).

Chemical information is defined here as data known to be about chemical substances. Database systems which handle chemical information have some special requirements, advantages and disadvantages when compared to conventional information systems. The most important advantage is that there is a universally-accepted model for the fundamental identity of entities for which information is known — the molecular structure. Conversely, the most important disadvantage is that essentially all known chemical information is ambiguous with respect to molecular structure. The reason for this is that chemical information is obtained for 'real' substances which generally correspond to more than one theoretical molecular structure. For instance, a measured boiling point is often measured not for a 'pure' substance, but for the constant-boiling mixture of isotopes, isomers and enantiomers which are in the pot being boiled. At the same time, data must be stored about very specific entities, e.g. an energy calculated for a specific conformation. To be effective, general chemical information systems must deal with this intrinsic diversity of data.

The importance of the molecular model to chemical information imposes special requirements on systems which are designed to handle such data. In particular, there are a number of access modes and search types which are specific to molecular structure data, e.g. data must be retrievable from partial specifications of tautomeric, substructure, superstructure and structural similarity.

Chemical information systems are special with respect to a number of other characteristics. Compared to most conventional database systems, chemical data systems are characterised by a very high degree of heterogeneity. Thousands of types of data are known about chemical entities, each of which is of critical interest to some users (e.g. representing a chemist's lifetime work) but all of which are of interest to almost no one (except possibly the system manager). This characteristic favours distributed, rather than centralised, database entry, maintenance and quality assurance. The hardware environment of real-world chemical database systems is also heterogeneous. Whereas systems designed for drivers-licence or airline reservation data can be designed for a single user interface delivered on homogeneous hardware, chemical information systems typically operate in very diverse environments. Users of such systems are as diverse as the hardware they use, e.g. business and marketing (mainframes), stockroom clerks (PCs), data entry clerks (low end workstations), synthetic chemists (Macs), modellers (high-end graphic workstations), computational chemists (multiprocessor servers), analytical chemists (processors dedicated on a single instrument) and students (anything they can get their hands on). To operate effectively in such an environment, chemical information systems must possess a high degree of hardware and operating-system independence as well as maintaining a high degree of representational generality.

Structure-based databases

Since the universal language of chemistry is the molecular structure, the molecular structure itself is uniquely suitable for use as the primary key for entries in a chemical information database. A system of structure-keyed databases has the advantage of intrinsic compatibility, i.e. all databases use the same key (structure) with no need for assignment of arbitrary indexes.

Unfortunately, a molecular structure *per se* is not a suitable index for conventional relational- or dictionary-oriented database systems. Conventional systems use a 'registration number' (or index) as a key and store structural information as data. Registration numbers generally contain no information about the structure (they are information-content-free). Rationalisation of a registration number with structure is typically done by a separate process known as 'registration'. Such systems suffer from two fundamental deficiencies: databases are not intercompatible (there is no enforced relationship between registration numbers for the same entity in different databases) and the registration process can support only a single arbitrary mapping of chemical entities to keys.

The SMILES nomenclature has several critical features which make it suitable for use as a structural database key [D. Weininger, 1988]. The most important one is that a unique SMILES can be machine-generated. Other important features for databases purposes are that it is a comprehensive nomenclature, it can be parsed easily, it is portable (uses only standard ASCII characters) and is relatively compact. These requirements are not met by conventional connection table representations nor linear notations such as WLN (Wisswesser Line Notation) and Rosdal.

One important characteristic of SMILES-keyed databases is that storage and retrieval may be accomplished in constant-time, i.e. the amount of time required to access data keyed by SMILES is not dependent on the number of entries in the database. This behaviour is obtained by treating the unique SMILES as the 'address' of the entry using the hashing method (Knuth, 1984).

Thesaurus-oriented retrieval and data archival

As noted above, chemical information is known about a wide variety of identifiers. With respect to databases, the chemical nomenclature problem is how to effectively describe what substance(s) data are known to be about. This problem is not due to nomenclature inadequacies. In fact, chemical identifiers are intrinsically overlapping and ambiguous. For instance, observations of the boiling point of 'dichloroethene', '1,2-dichloroethene', 'cis-1,2-dichloroethene' and '37Cl-1,1-dichloroethene' are all valid but require different identifiers. The fundamental paradigm of computerised databases is the dictionary, which basically provides the capability to store one or more datum associated with a given entry. Relational databases do not solve this problem because the relationships that they describe are between types of data rather than data items themselves. Neither method is suitable for effectively storing data keyed to ambiguous identifiers.

The thesaurus is a more suitable paradigm for storage of data about overlapping and ambiguous entities. Thesaurus-oriented retrieval (THOR) keys data to 'topics' while maintaining the identifier which each datum is known to be about. In Daylight's chemical information databases, THOR uses the unique SMILES of the hydrogen suppressed graph as the thesaurus 'topic'. Identifiers may be more specific than this key (resulting in multiple entries per topic, e.g. '37Cl-1,1-dichloroethene') or less specific (ambiguous with respect to structure, i.e. 'dichloroethene').

A powerful characteristic of THOR databases is that data can be accessed by all possible types of identifiers in constant time. This is done by accessing the unique SMILES by 'hashing' the identifier, then accessing the data keyed to that SMILES as previously described. All THOR retrievals, whether accessed by SMILES, name, registration number or whatever, are accomplished in three or four disk accesses, regardless of database size or complexity. Furthermore, all THOR databases are fundamentally inter-compatible with respect to all possible identifiers. THOR effectively solves the chemical nomenclature problem with an algorithm which operates in constant time.

The scope of constant-time-access-identifiers extends beyond data types like SMILES, names and registration numbers. Any unique identifier of structure may be used to provide direct access to data. For example, the 'graph' of a molecule is the unique SMILES of the hydrogen-suppressed structure with all oxidation state information removed. The Daylight implementation of THOR provides automatic graph generation on database load, allowing retrieval of all structures with the same graph in a single lookup. Structures with the same graph, hydrogen count and net charge are considered putative 'tautomers'. Thus, all known data for tautomers can be retrieved from a THOR database in constant time.

THOR also allows storage of chemical information for which a structure is not known. Such information is particularly important in analytical chemistry applications, where data are observed for substances with identifiers like 'Unknown 1344'. Data are keyed to the given identifier and THOR becomes essentially a dictionary algorithm. If a structure is later assigned to that identifier, the identifier and all associated data are moved to their proper structure-keyed place in the thesaurus.

In-memory exploratory data analysis

The primary role of searching in modern chemical information systems is exploratory data analysis (EDA), i.e., looking for and characterising information in a database without previous knowledge of what data are available. For such purposes it is critically important that search methods be accurate, flexible and fast. Fortunately, the amount of data in chemical information databases can be stored in core memory, even for very large databases! For instance, SMILES average about 25 characters in length for most databases, which means that 1,000,000 structures can be stored in 25 MB of core. Storing searchable data in-memory is feasible because current generation machines have high memory capacities, e.g., 64–256 MB on workstations and 512 MB–4 GB on servers. Searching large databases in-memory does require a significant amount of computational machinery, e.g. about 400 MB are required to load the entire Spresi93 database which contains 2.2 million structures, 1.1 million physical properties, 3.5 million journal articles and patents. Search times for this database take 1 to 30 seconds, depending on search type and server hardware.

In-memory storage has a 100,000-fold performance advantage over disk-based systems for data access. This phenomenal increase in access speed allows implementation of flexible search methods which are suitable to EDA. Searching for similar structures can be provided along with traditional substructure methods. Searching text data can be done via free text searching or with generalised regular expressions rather than with keywords. In-memory sorting is fast enough that sorting the whole database by a desired data type becomes preferable to entry selection based on range. In general, database searching problems no longer revolve around getting data into and out of a computer, but rather how to provide the user with a query language which delivers data relevant to a particular task.

Implementation of network servers

The Daylight implementation of THOR uses a disk-based TCP server. A single server, operating on a 'host' machine, delivers database services to client programs on a network. Use of a database server allows simple and effective methods of providing database security (the server acts as a gatekeeper), data integrity (only the server program accesses archival data), multiple read/write-access (TCP provides collision resolution) and resource utilisation (a high-performance server can deliver services to many clients). THOR databases are ideally suited to distributed database maintenance. Given a suitable network, THOR databases are equally accessible in a local area, a wide area and globally.

The Daylight in-memory search engine is named 'Merlin' and is also implemented as a TCP network server. The client-server model is eminently suitable for this purpose, since the Merlin server typically uses a lot of memory, an expensive resource which should be shared. Merlin uses the same data as the THOR database server, i.e. in principle, any data in a THOR database can be searched by Merlin. In practice, only a subset of THOR data is selected for loading by the Merlin search server, since much of it is not interesting for EDA (e.g., time-stamps, graphs, cross-references, etc.)

A third TCP server is the Daylight Remote Toolkit Server, which provides object-oriented chemical information services to the network, such as unique SMILES generation and structural depiction. The remote toolkit server provides capabilities which may be used in client applications running on less capable machines such as personal computers and special-purpose processors dedicated to analytical instruments.

Implementation of high-performance in-house client applications

Database services described above are strictly network services which are delivered in a way which is completely independent of hardware, operating system or user interface. For in-house use, it is natural to mate the high-performance capabilities of these database servers with high-performance user interfaces. Given the availability of TCP networks operating at relatively high speed (e.g. Ethernet), the preferred high-performance user environment is currently a Unix workstation operating X-Windows. X-terminals and X-terminal emulators also provide high-performance user-environments. Separate client programs oriented to THOR and Merlin have been implemented and are available from Daylight.

The *xvthor* application program provides a 'microscopic' view of THOR databases. The fundamental unit of operation is a data tree, e.g., one SMILES-rooted entry with all associated identifiers and data. *xvthor* is designed to allow the user to examine database entries in detail, examine identifier relationships and enter and edit all types of data (including adding new data types).

The *xvmerlin* application program provides a 'macroscopic' view of THOR databases and is designed to facilitate exploratory data analysis. *xvmerlin*'s fundamental unit of operation is a database, which it displays as a spreadsheet, with each row representing an entry. Columns represent types of data, to which various functions may be applied, e.g. 'first', 'average', 'shortest', etc. The entire database may be sorted by the value of data in any column. Rows may be selected in a very flexible manner based on string value, numeric value, ranges, structural similarity and presence of superstructures or substructures. Graphical depictions of structures are integrated into the spreadsheet. 3-D data are displayed in a trackball-based window. *xvmerlin* is also a client of the THOR server and allows detailed data display, similar to that in *xvthor*, but on a read-only basis.

A number of other programs are provided for database control and management. These programs use a simple serial user interface. The most important such program is *sthorman*, which provides management capabilities such as creating a new database, authorising access, allocating search pools and monitoring usage.

Implementation of medium-performance clients

Many potential users of chemical information systems do not require fully-generalised database access and typically use limited-performance machines, such as personal computers. Since most companies include such machines in their networks, it is reasonable to extend chemical information access to such users.

The existence of reliable, object-oriented, networked servers greatly simplifies the task of incorporating database services into special-purpose programs. Since the servers guarantee data security and integrity, adding database access to a program amounts to little more than providing the user a means of asking for it.

Savant, a literature searching program for synthetic chemists, exemplifies such an application. *Savant* is a 'double-clickable' Macintosh program which is a client of all three Daylight servers and accesses the Spresi database which contains journal articles describing the synthesis of over one million structures. Users enter the SMILES of a structure of interest, either manually or via a program such as ChemDraw or Chemintosh. *Savant* selects journal articles in the Spresi database with the keyword 'preparation', sorts them by similarity to the given structure (using the Merlin server), draws the most similar structures (using the Remote Toolkit server), retrieves title, authors and full citation (using the THOR server) and displays them in a window similar to the Macintosh Scrapbook. The resulting program provides a user-friendly literature search for syntheses of interest. The Macintosh client is very lightweight because all the real work is done externally (it is a 100K application).

A number of third-party vendors are developing chemical database programs using Daylight servers. Several pharmaceutical companies are doing the same thing for their own in-house use. Typical target hardware for such programs is a Macintosh or IBM PC.

Implementation of low-performance data services

A final class of chemical information users operate over low-performance communication mechanisms, e.g. academic users accessing public databases via e-mail. The archetype for such service is *mjollnir*, a simple chemical information access program supported by Daylight which operates exclusively over an e-mail interface. Users simply send their request as a mail message to *mjollnir* and receive an e-mail reply. Queries are specified on the mail subject line, e.g. 'getdata' and input data appear in the message body, e.g. identifiers such as SMILES or names for which data are desired. The *mjollnir* system is low-performance because the response time is slow, e.g. 2 to 12 hours.

Since the through-put of mail is relatively slow, its impact on fully-operational database servers is negligible. As a consequence, Daylight is able to supply this service free of charge (*mjollnir* responds only to educational addresses).

Although *mjollnir* is now used as a special-purpose program for supplying chemical information to academic users, it may have a potentially large impact on chemical information delivery. *Mjollnir* provides a mechanism for publishing data which is more effective than conventional mechanisms in almost every respect. Anyone with access to e-mail becomes a potential user. All high-performance chemical information system attributes apply to *mjollnir* except speed, e.g. all databases are inter-compatible, any type of data may be stored, it scales up to huge databases, retrieval speed is independent of database size, etc.

Performance and reliability

The THOR/Merlin system has been used and tested at Daylight for several years on a wide variety of databases including: Spresi (bibliographic data and physical properties for 2.2 million structures), Beilstein Current Facts (about 350,000 structures and properties), Chemical Abstract's CAS3D (about 500,000 structures), Derwent's World Drug Index (pharmacological and trade name data for about 40,000 structures), Medchem (LogP and pKa data and references for about 30,000 structures), TSCA (U.S. Toxic Substance Control Act structures), Maybridge (about 35,000 structures of medicinal and agrochemical interest available for purchase) and others. Also, a number of companies are using this system with corporate databases containing 20,000 to 450,000 structures.

The THOR database system operated without malfunction throughout this period. THOR is designed to provide high-reliability data archiving and to be able to reconstruct a database even if it should become corrupted due to a power outage. Although there have been many system crashes on machines running THOR servers, THOR databases have never been corrupted to our knowledge. (This also means that the database reconstruction procedures have not yet been used in a real life situation.) The primary difficulty with using THOR has been in design of appropriate data types, a design problem common to all databases.

THOR database intercompatibility has proved enormously valuable. For instance, the Spresi database contains huge amounts of bibliographic, patent and physical property data, but no identifiers other than the Spresi Registration Number, which limits its usefulness for most users. The World Drug Index contains an extremely large number of names (includes all drug trade names, worldwide) but no biblio-graphic information. Using the THOR system, these databases become compatible despite the fact that the Soviet Academy of Sciences and Derwent did not agree on identifiers when building the databases. Users can look up a drug by name and immediately obtain literature and patent information. Corporate users can access the same data by specifying their corporate registration number.

The Merlin search system has operated well, but not entirely without difficulties. By far the most common problem has been running out of virtual memory. The Merlin server uses much more memory than most programs and many system managers are reluctant to allocate sufficient swap space. This problem is compounded because some operating systems are not designed to handle this situation well — Irix panics and blows away processes at random and HP U/X just becomes confused (to be fair, HP has a fix in progress). The current version of the Merlin server has a capability to limit the absolute amount of memory used, which has alleviated this problem. The best solution is to provide the server with enough real memory to do its job and to dedicate the machine to database services. This solution has not been very popular so far.

Merlin performance has been excellent, with the exception of a couple of unusual searches which needed to be re-coded for better performance on large databases. The availability of shared-memory multi-processors such as the Sun 2000 and SGI Challenge is providing scalable database services at several companies. As the user load increases with time, processors are added to provide the desired performance.

Database performance over a TCP network has been outstanding. The best example is a company which established duplicate database server facilities in both its US and UK sites, presumably to provide higher performance locally. Users discovered that they could access the remote site, since the company network serves both sites. Users now regularly access servers at the remote site in preference to their own because the extraneous load on that machine is lower, e.g., the British host machine is idle when American users need it due to the time zone difference. In this case a more effective solution might be to dedicate one machine to database service. None the less, this serves as an excellent example of how networked databases allow computer resources to be shared.

Conclusions

The system described here appears to be a chemical information system capable of delivering services to essentially all classes of users. This system is effective for small project databases and scales up well to large corporate databases. One of its primary advantages is that of database intercompatibility, which allows integrated access to a wide variety of data. End-user applications are limited to general-purpose programs more suited to the information specialist than the bench chemist or stockroom clerk, but this may change as third-party application software becomes available. Excellent performance is obtainable using currently-available hardware from a variety of vendors.

References

'CSC ChemDraw', Cambridge Scientific Computing, Inc., Cambridge, MA (1992)

'Daylight CIS Systems Administration', A. M. Weininger and C. A. James, Daylight CIS, Inc., Irvine, CA (1993)

'Daylight Toolkit Programmers Guide', C. A. James, D. Weininger and J. Scofield, Daylight CIS, Inc., Irvine, CA (1993)

'SMILES, a chemical language and information system. I. Introduction to methodology and encoding rules', D. Weininger, *J. Chem. Info. Sci.*, **28**, 31 (1988)

'Sorting and Searching', D. E. Knuth, Adison-Wesley, Reading, MA (1973).

'Infobit Retrieval': assessing the correct delivery technologies for chemical information

Randall Marcinko

Dynamic Information Corp., Burlingame, California, USA

Introduction

The information industry has grown dramatically. As with any fledgeling industry, growth has been uncontrolled, often in opposing directions and with uncertain goals. A plethora of resources, techniques and products are available to today's chemist. There are libraries, information specialists, information scientists, information consultants and information professionals all seeking out the chemist.

If each information-rich item needed by the researcher is termed an *Infobit*, there are a host of products and providers that deliver *Infobits* to the chemist. The information broker is that *missing link in the information chain* that can function as a facilitator for information retrieval. The last two years have seen a great interest in document delivery, not only by chemists, but universally. The information broker does not create documents, nor does he create a need for them. The broker is a conduit that gathers the desired materials, funnels them through a central location and delivers them to the client. The document is a commodity, just like a chromatography column or an online search. It is a 'bit of information' that enters the research process as does any other raw material. The Infobit can be paper copy or electronic. It can be a product sample, a prior art search, an expert contact, or any other information-rich item. The information broker serves as a conduit to gather and deliver Infobits. It is the broker who has long been overlooked as an intermediary facilitating information transfer.

Traditionally, the chemist has viewed the librarian as a technician who slowly and inefficiently retrieves information. The information broker has never commanded the same respect as the materials scientist who assists the chemist in product testing. Nevertheless, Infobit Retrieval is a pursuit that has started to mature.

Assessment of the correct delivery technology can only be realised through communication between the retrieval expert and the chemist/end-user. ARIEL and T1 connections, and ADONIS and Kofax cards are new developments that are readily available, but, does the chemist really want megabit/second transmission of bitmap images via the Internet? How important is access to melting point data at midnight or a specification sheet to better use a new GC column?

Historically

The chemist has long depended on information for the successful performance of theoretical, experimental or analytical research. Chemistry is without question one of the most precedent-oriented of the sciences. In 1900, Moses Gomberg postulated the existence of the triphenylmethyl radical, however, his peers were slow to accept the heretofore undocumented species. How did Gomberg's colleagues know about his research? They relied on word of mouth, on meetings among peers, and on reading the scientific journals and reviews of the day. While information science has grown rapidly, the chemist has not and is not taking significant advantage of the world of information. Unfortunately, while databases, online access and molecular modelling are now a reality, much of the information needs of the average chemist remain unanswered by the information professional. The sources and resources exist. Why is the chemist not taking proper advantage?

The chemist understands the need for the biologist to conduct bioactivity testing. Why can the same not be said about the need for an information scientist? Three factors need to be addressed:

- *The chemist had long been the leader in information retrieval*

For decades before the early online databases were designed and released, the chemist had been indexing articles, reactions, and other facts by punching holes in the sides of 6x8 cards. Retrieval seemed crude as the chemist executed search statement after search statement by inserting rods in particular positions along the edges of the cards, often stored in shoe boxes. Those cards that 'fell out of a search' were the gems of information. Abstracts on each card furnished added data and usually pointed to the original literature. As groups of chemists shared cards and standardised indexing, these shoe box databases grew in size and in comprehensiveness. How did the chemist react to the first online databases? While the librarian of the day was elated by the early searches on a teletype terminal, the chemist often returned to shelves of shoe boxes where the journal coverage and quality of abstracting and indexing were superior. Out of a need for good information, the chemist had a very reliable, albeit somewhat labour intensive, database engine. When the librarian/information scientist said to the chemist, "Look what we have done for you! We have created a database for which you can pay money and from which you will have answers." The chemist asked "find all of the molecules with a sulphur group and which were made using a Diels-Alder reaction" since these cards certainly existed in the shoe boxes. However, the answer came: "what are some good keywords? You don't expect structure searching, do you?"

The information industry has matured. However, the chemist remains reluctant to pass responsibility on to the librarians/information scientists who feel they were first to design chemical databases. Perhaps 'chemist' is synonymous with conservative. Precedent is, after all, one of the first words learned by budding chemists.

- *The chemist traditionally understands the qualifications of colleagues in other scientific disciplines*

If a chemist produces a new molecule with potential biological activity, it is typically handed over to a biologist or medical researcher for testing. The chemist wishing to look at the photoelectron spectrum of a class of compounds readily collaborates with a physicist. However, the chemist places much less trust in the information specialist. Is this well-founded?

Several parameters must be evaluated to answer this question. The biologist and physicist are important because they can perform tasks that the chemist would be hard pressed to duplicate without a long learning process. The assumed experience of scientific colleagues lends credibility and reliability to their results. Until the chemist is aware of the special capabilities of the information scientist, trust and reliance on outside information support will be limited. Until the results are consistent, more quickly obtained and of higher quality than the chemist's own searching, the information scientist will be relegated to the role of junior technician. With the vast array of resources and techniques, it is incumbent on the information scientist to reach out and show the chemist what he is missing. The information scientist must find those resources that are truly needed by the chemist and which are lacking in the workplace.

- *The information scientist, the librarian, the information specialist, the para-professional and others in the information industry present a confusing front to the chemist*

The small graduate college with limited funds may not be able to give its students unlimited access to online searching — they will certainly not be able to give access to the most advanced molecular modelling programs. The most highly funded information centres in pharmaceutical companies will provide molecular modelling systems, limitless online access and a wide array of electronically stored spectra for analysis and comparison. But does the information centre obtain product samples? Both the student and the professional will need to browse printed manuals, probably make numerous telephone calls and quite likely conduct manual library research before locating a product sample or the manufacturer of an uncommon substrate. Are the information needs of the chemist being met?

The answer to this question is much more positive than it was twenty-five years ago. Nevertheless it is still *"no"*. The information industry is inconsistent in its quality and in its presentation and image. This results in low use and low trust on the part of the end-user/chemist.

Needs of the chemist

While similar analogies exist in all chemical disciplines, for the purposes of illustration, this paper will consider the organic bench chemist and the typical questions and problems that must be solved. For decades, organic chemists have been asking the same questions. How many new solutions has the information industry provided? How many solutions exist and are not being put into the hands of the chemist? The answers to these questions point directly to the missing link in the information chain.

- before embarking on a new project, how do I search the literature back to 1970?
- before embarking on a new project, how do I search the literature before 1970?
- how do I obtain document delivery to support the online searching?
- before embarking on a new project, how do I find out who, if anyone, is currently working on the same subject area?
- how can I thoroughly analyse the patent literature in the desired product area?
- before embarking on a new project, how do I locate experts in related fields?
- how can I locate topical seminars, conferences and workshops?
- how can I locate additional funding sources for my research?
- how do I apply molecular modelling and predictive reaction technology to my project?
- how do I locate commercially available starting materials?
- how do I purchase commercially available starting materials?
- how can I locate research samples of products similar to my target molecules?

- how can I obtain information on the physical properties of my starting materials?
- how can I locate nmr/ftir/ms spectra of starting materials and products?
- how can I locate labs that can run spectra or take other physical measurements on my products?
- how can I communicate with others in the same research area and share information during the research process?
- how can others be made aware of my research in progress so that I might discuss ideas and possible solutions?
- how can I find information on the environmental impact of my research?
- how can I locate manufacturers for the special equipment that I need?
- how can I purchase the necessary equipment?
- how can I best publish the results of my research?
- how can I locate marketing information on the product and on related materials?
- how can I find names and contacts of others potentially interested in the products of my research?

Such questions are typically asked by organic bench chemists in the course of their research. While the information industry can provide many high-quality answers, the chemist is still very often unaware of these possibilities. Many information professionals ignore the importance of the apparently 'trivial' questions. The information professionals *must* be aware of all concerns of their patron constituency and must have a solution for each. Does the information professional feel that answering some of these questions is too menial a task? Is the information professional unaware that many of these questions are vital?

The information broker is able to facilitate the process. Each Infobit addressed in the above list is a commodity that can be located, gathered and delivered. While the information professional is able to work with the end-user/chemist in solving problems, the information broker is the Missing Link in the Information Chain. The broker functions as a conduit to be used by the professional and end-user alike for the delivery of information. Figure 1 is a pictorial representation of some of the information needs facing the organic chemist, conducting research on a new chemical reaction.

The information world

The information-illiterate chemist should be able to demand information on any item portrayed in Figure 1. The information professional is there to assist the chemist in navigating the sea of information. The professional should guide the chemist by answering questions and should be proactive in offering support that will avoid problems down the line. Today, most information providers perceive themselves as experts in specific areas. They perceive certain functions as menial and yet feel they have to be able to handle each task personally. As in every profession, specialisation occurs. However, the leaders in each profession are those who can manage and direct a team that includes many subject experts. Within each successful research team are connections to the outside world, to the world of related disciplines. Analogously, the information broker is positioned to assist in the quest to obtain the desired Infobits. The broker is not the information professional who guides and navigates the corporate researcher through the information sea, and who understands all of the separate missions of the corporation. The broker is the conduit that has special expertise in the locating and retrieving Infobits.

Literature Searching (pre-1970)
Literature Searching (post-1970)
Prior Art Patent Searching
Document Delivery
Locate Subject Experts in the Same Field
Locate Topical Seminars, Conferences and Workshops
Employ Molecular Modelling

STARTING MATERIALS

locate reaction vessels
locate research quantity product samples
locate chemical sources
locate spectral information
make starting material purchases
locate physical property information
locate equipment manufacturers

Environmental Concerns
Political Concerns
Economic Concerns
Market Considerations

Communication With Other Researchers
Monitor Related Research in Progress

PRODUCTS

Locate Marketing Information
Publication of Research Results
Locate Others Interested in the Product

Figure 1. 'Information needs surrounding a new chemical reaction'

Infobit Retrieval

The information needs of a pharmaceutical research team are analysed. The researchers are charged with a typical task:

TASK: Company X needs a method to produce a large quantity of organometallic substrate Q. Company X wishes to use the material in the production of a new antibiotic. Company X also wishes to market the organometallic substrate as a separate product. The research team must find a profitable method of production and test the process at the pilot plant.

The information professional and the research team meet initially to discuss the project. Both try to identify needs: the information professional tries to supply the best possible solutions. The information professional responds to the needs of the research team.

To obtain the Infobits, an information broker can be used in a variety of ways.

Document delivery: Document delivery, has come of age. Over the past two years everyone has been clamouring to obtain the elusive document. While libraries have been the traditional repositories of books, monographs and journals, the slashing of budgets and explosion of published sources reduces the utility of any individual library. Through the intermediary of a broker, any document can be obtained in the desired time frame.

Brokers use international library collections, relationships with publishers, relationships with database producers and other partnerships to broaden their sources. Virtually any article can be obtained. Delivery of documents can be by mail, by express courier, by fax or even by Internet fax. It is essential that the necessary price and time constraints are declared up front. Documents can be ordered from any location via telephone, fax, mail and the Internet.

Literature searching (pre-1970): The information professional or the end-user can use a broker to have manual searching conducted for periods before 1970. This might be deemed necessary when the end-user or the library does not have a specialist with the correct subject background. Perhaps a polymer chemist is needed to do an in-depth search on a complex topic.

The end-user and information professional are often unaware of the scope of manual research that can be handled by information brokers. Brokers employ telephone research, library research and use resources from all over the world. If a substrate were first synthesised in Japan in the early 1950s, the broker can have a Japanese researcher retrieve data from a local Japanese pharmaceutical library. If one of the original researchers were in Germany, the broker could use a German speaker to contact a scientist in Germany.

Even more important, a report could be prepared that summarises and digests the available information. The report could be as comprehensive as to include interviews and transcripts of conversations with industry experts. The Infobits being retrieved are important facts from the past. They can be delivered in raw or report form and can be accompanied by peer comment. The degree to which the information should be digested and assimilated is determined by time, budget and end-user needs.

Literature searching (post-1970): The information professional or the end-user can employ the broker for online and manual searching conducted post-1970. This might be deemed necessary when the end-user or information centre does not have a specialist with the correct subject background. It may also be needed when online searching is to be conducted on a system not held in-house or for which no trained staff are available. The chemist need not be denied a search on any online system or database, just because a library does not regularly search on the database or network.

The broker can use experts on any subject or on any vendor, host or database. The information professional need not be intimately familiar with the subject to retrieve the best data available via the broker.

Prior art searching: The broker can be used for patent searching when in-house searchers do not have the proper subject background or are not patent experts. Even the researcher in a small corporation or from academia should be able to benefit from comprehensive, high-quality patent searching when these Infobits are obtained through a broker. The broker's product can include patent copies, online search results, manual searching conducted in worldwide patent offices and interviews with experts skilled in the art. Interview and transcript information can be included from conversations with patent experts. Information on competing products and services will often be an integral part of the information product.

Product samples: Just as documents are Infobits, so are product samples. The chemist working on a new additive for yoghurt may need numerous competitive samples from around the world. The information broker has the ability to import samples from any country and to deal with appropriate customs regulations. Most end-users and information professionals will not encounter such instances suffi-ciently frequently to obtain products quickly and efficiently. Moreover, the broker can obtain the material completely anonymously.

Experts, conferences and workshops: Perhaps a researcher would like some brief consultation or an opinion from an expert. Perhaps the researcher would like to attend topical meetings with peers. The broker is able to seek out people and conferences, anonymously where necessary.

Molecular modelling: Even the small company or academic researcher without large in-house systems, can take full advantage of capital-intensive resources. The problem can be placed in the hands of a broker who can use the hardware and software of a third party upon payment of appropriate fees and licenses.

Chemical sources and starting material purchases: For many chemicals it is extremely arduous to locate or even to ascertain whether a commercial manufacturer exists. The chemist, the purchasing agent and the information professional are often not the most useful in locating the most cost-effective source. The small company, that does not often import chemicals from around the world, will find the interme-diacy of a broker invaluable in avoiding customs and import/export problems.

Research quantity product samples: The chemist who wishes to obtain small amounts of research chemicals may spend many hours on the telephone locating milligram quantities of a material that another chemist has synthesised. The infor-mation professional can treat these materials as Infobits. Information brokers are capable of tracking down and procuring such materials. The broker may be vital when anonymity is desired or when foreign languages or import/export regulations become an issue.

Spectral or physical property information: Not only are starting materials impor-tant, the proper spectra and physical properties can take a great deal of time to obtain for comparison purposes. Whether from the chemical manufacturer or from another researcher who previously synthesised the material, these Infobits are in the domain of the broker. No information professional should neglect to assist the chemist in the procurement of these Infobits. Spectra, product samples and melting points are all valuable pieces of information routinely requested by the research chemist.

Equipment manufacturers: "I am having difficulty separating my products!" "I cannot locate the proper chromatography column!" These are all-too-common statements in the chemical laboratory, but rarely are they made to the information professional. The solutions are Infobits that information brokers are able to obtain. Other scientists who have worked with similar compounds can be contacted. Chromatography equipment manufacturers can be contacted. The broker can deliver the information in summary report with a sample column able to successfully separate the materials. The broker can even contract a third-party laboratory able to handle the entire separation process.

Environmental considerations: The chemist may need to evaluate the environmental impact of waste disposal, or evaluate the fire hazards when handling a particular compound. The broker can efficiently provide these Infobits through online or manual searching, through Freedom of Information inquiries or through interviews with experts and government officials. The broker may be the source of choice when anonymity is a concern or when the end-user/information professional is not an expert in environmental research.

Communication with other researchers: The chemist often wishes to communicate with peers working in the same area or to monitor the research of others in the field. It may require significant effort for the chemist to track down Internet or other sources for this data. The information professional should be prepared to supply electronic access points and to prepare lists of experts. The broker is able to deliver such Infobits. The information may be compiled using printed reports, using online reports on through telephone interviews and in-person contact with experts in the field, either in their laboratories or at industry gatherings. The chemist or information professional may choose to have the broker attend and report back on pertinent information being delivered at various trade shows and subject workshops. The chemist may choose to use the broker to contact peers on the Internet and discuss research in progress or to post information for question and/or comment.

While these are only some of the possible Infobit retrieval tasks that the chemist or information professional can delegate to the broker, they are a bare minimum that should be considered in this example. The chemist must demand at least this level of information support. The information broker is often the *missing link* that can facilitate the process.

Conclusion

The world of the chemical information professional is as complex as the variety of problems conceived of by the practising chemist. Because of the information-oriented history of chemistry and unbridled development of the information industry, the chemist often views the information professional as inferior technical support. The information professional often concentrates on only a small part of the overall information needs of the chemist. For the chemist to elevate the information professional and information industry to a level on par with other academic colleagues and disciplines, all of the needs of the chemist must be met. The professional should command his ship as a captain and navigate through the available information sea. On this journey the corporation, the researchers and all parties involved should have access to the information they require.

The information professional is not alone in the provision of each bit of information. Rather the professional has ready access to information brokers. The broker is able to deliver any Infobit to the end-user or to the intermediary. The broker specialises in locating, gathering and delivering the Infobits. The information broker is useful as that often missing link in the information chain.

Chemists should demand access at least as comprehensive as alluded to herein. They should demand to work with information professionals who can assist them with this task. Information professionals are truly colleagues who can be folded into the research team. Other providers should be viewed and employed as technicians. The information broker is available to the chemist or to the information professional/ intermediary. No Infobit is too elusive to be obtained. Once the chemist, either individually or in concert with a professional, ascertains needs, the broker can obtain the Infobits. The information professional orchestrates the process. The chemist must demand quality — it *is* available in today's information world.

Today's information, tomorrow's technology

Elizabeth M. Hearle

MicroPatent, Cambridge Place, Cambridge CB2 1NR, England

Introduction

Whilst online and CD-ROM continue to dominate the searching and hard-copy delivery of patent documents, both systems have up until now suffered indisputable drawbacks.

Current online patent information systems allow text-only searching and do not provide a means of examining the complete patent document in its original format. Off-line systems such as CD-ROM, require complex jukebox arrangements to effectively handle the large number of discs required for just a few years worth of patent documents.

MicroPatent has addressed both these problems, and this paper describes our new approach to the search and delivery of patent information world-wide. Our objective is simple: to make patents available to everyone in an easy-to-use format, combining the very latest in both search and document delivery capabilities.

The proposed system will enable the searching and full-text printing of customised patent databases; this will by-pass the problems of high online access costs or increasing collections of discs containing a high proportion of irrelevant information.

Initially, the system is being used for the development of custom in-house patent databases for both large and small companies. By next year, the system will evolve into a complete online service.

This could radically change the way patent information is currently being used.

The system in question? PatentNotes

PatentNotes is a customised in-house patent library especially designed for companies that already have an existing computer network. PatentNotes is designed to sit on the existing network and via a common graphical user interface allow a user to access, download, display and print selected patent information, including both text and images; a minimum of computer skills is required.

The system involves the use of a dedicated patent server on which the customised patents reside. The server communicates with its users through the existing network operating system. A number of systems are supported, including NetWare, LAN Server, LAN Manager, Banyan, Pathworks, Appletalk, Windows for workgroups and TCP/IP. Even remote users can access the server via modem and telephone line.

The server can be anything from a 486 PC running Windows to an OS/2 or UNIX workstation, available from suppliers such as IBM, Hewlett Packard and Sun Microsystems. The server currently supports multiple clients such as Windows,

Mac, OS/2 and AIX. It is also expandable from a small local area network (LAN) involving thousands of users.

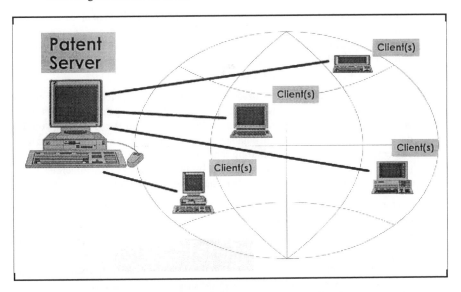

Figure 1 shows how this 'electronic patent library' works. A dedicated patent server running under Lotus Notes contains the fulltext/image database of selected relevant patents; searchable front-page text may alternatively be substituted for full text. Users choose which patents to include in their customised set. Patents may additionally be arranged into one or more sets based on a broad number of criteria, including industry, patent classification, company names, key words, date range, etc.

As an example, all US biotechnology patents may be placed under a Biotech icon, chemical under a Chemical icon, etc. Each database also contains the scanned-in images of the complete patent for document delivery. Data housed in the server are available with unlimited access to any or all parts of the organisation incurring no access charges.

At the present time, both US and PCT patents are included in the database. MicroPatent is currently negotiating with the European patent bodies to add these patents as well.

The database is initially loaded on to the server and updated from a central location. Updates can be performed at regular intervals (weekly) or on demand.

Why Lotus Notes?

Lotus Notes was selected as the retrieval software because of its position in the corporate community as an advanced workgroup software with E-mail and shared-document database capabilities. In the last three years Notes has established itself as a strategic tool for workflow management in banks, insurance companies, legal firms, chemical and pharmaceutical companies, utilities and manufacturing organ-isations. As a database management tool, Notes also provides excellent search and imaging capabilities. The search engine provides for all the latest term and Boolean capabilities plus relevancy checking.

There is also considerable third-party software that works within or alongside Notes to collect, combine and automatically monitor information from a wide range of file sources, including patents. Through this software it is possible to completely automate the collection of pertinent patent information for delivery to other data-bases, E-mail, monthly reports, etc.

To the typical corporate user of Notes, the addition of patents is just another on-screen file folder. Opening the folder reveals the various patent databases, each with their own distinctive graphic icon. Clicking twice on the icon brings up the database. Once inside the patent database, for example containing US patents, a user can employ a powerful set of finding tools, including term and Boolean searching to isolate the patents of interest. Patents can be viewed and sorted in numerous ways: by company, inventors, patent numbers and dates, for example.

Searching across multiple patent databases or a patent database in conjunction with another database is similarly possible.

Notes' replication feature enables users to copy all or part of the database to a local drive and sort it in a variety of ways, copying sections to file or other applications, and storing search strategies to files for re-use. Standard network security precautions in Lotus Notes prevent changes to the original server database and establish who on the network has authorisation to access the data.

Search and view features

PatentNotes has all the features of a powerful search engine. A window appears upon entering the database, allowing the user to look up patents by 15 fields or field combinations, including patent number, issue date, assignee, inventor(s), patent class, title and abstract. Users can perform Boolean searches, proximity searches, wildcard searches, phrase searching, and numeric value searching on all titles, abstracts and/or full text.

Additionally, the full images of the patents are available for viewing or printing. These image files are electronically 'attached' to the patent file, and are selectable by clicking on the image icons. Using selected tools, one can perform enhanced functions such as zoom, rotation, cut and paste and selective printing of the desired patent images.

Images may be downloaded to file or exported to other applications. One or more pages of the patent document may also be faxed or mail merged.

A breakthrough in patent information

Now, patent information is no longer the sole province of the information specialist. Through PatentNotes, patent information is and must be a resource available to everyone in business. Many company projects today are cross-functional in nature, requiring collaboration among a wide range of groups, both internal and external. Notes' team-sharing capabilities allow patent information to be posted into a 'discussion database', along with memos, reports, meeting minutes, news wires and journal articles, for review and comment by other team members. Through the discussion databases, colleagues can brainstorm ideas, hold meetings, and solve problems from their desktop or laptop computers, breaking down the traditional barriers of geography and time zones. Information can then be extracted, cut and pasted with other data, such as graphs, charts, R and D data, memos and reports, then faxed or E-mailed to other parts of this organisation. It is a whole new way of empowering people by providing instant access to the latest technology from around the world.

Breakthrough at what price?

Any new technology begs the question, "what price?" Notes is surprisingly cost-effective for both the small and large corporate user. Lotus Notes starter packs are available for one server and two clients for less than $1,000. The modular nature of Notes also allows users to expand coverage as needed at approximately $500 per client. Once information is downloaded to the server, there are no further access charges. The information is available for unlimited usage.

Future developments

The client–server nature of Lotus Notes, with its open architecture and distributed shared databases under many network and hardware platforms, makes it ideal for building a variety of configurations. We anticipate a spectrum of different corporate users who want local servers configured to meet their specific information needs. High-end users may want local access to 'all' patent data sorted into selected categories, while medium-end users may want local access to only selected patents of specific interest areas. Low-end users may want dial-up access to an online network for an occasional patent. In fact, we expect PatentNotes to be the ideal document delivery tool for experienced online users.

In the future, patent information will command the same interest and attention as other information sources for tracking industry and competitive activities. Patent-Notes will facilitate fast, easy sharing of patent information by everyone, no matter where, when or how they work.

MicroPatent, the first company to deliver images of US patents on CD-ROM, is pleased to be the pioneer in this venture.

A market-driven approach to US patent data

Geoffrey Trotter

Research Publications International, P.O. Box 45, Reading RG1 8HF, England

The Thomson Corporation, one of the world's largest publishing conglomerates, has maintained its interest in the provision of patent information through three subsidiary companies: Derwent Publications (with value-added abstracts), Thomson & Thomson (Trademarks) and Research Publications International (full specifications of patents).

For over 25 years Research Publications International (RPI) has provided national patent offices, agents, attorneys and industrial corporations with a wealth of patent information in a variety of media — paper, microfiche, microfilm and online. Since the late 1980s, RPI has been involved with the research and development of a number of electronic databases. Production of CD-ROM databases started in 1990 and to date four CD-ROM products are being offered to a global customer-base via offices in the United States and the UK.

The databases are *PatentView, OG/Plus, PatentHistory* and *PatentView Backfile*. In addition to these four 'in-house' products, RPI is a distributor for the ESPACE series CD-ROM databases produced by the European Patent Office.

So what is it that RPI has done in its approach to bringing US patent data to the desks of many information professionals?

Focusing on PatentView, RPI's full-image US patent product, we will look at the ways in which this database can be used to provide information in a timely, convenient and flexible manner.

PatentView

PatentView is RPI's flagship product. It is a database of *all* US patents issued on a weekly basis and within two weeks of the date of issue of the patents by the USPTO. It is therefore an excellent current awareness tool.

The database is composed of two parts:

- Searchable, indexed bibliographic textual data
- Printable, viewable graphics images of the pages of the patents.

Delivered either in complete form, or in a standardised chemical, electrical or general/mechanical subset, users of PatentView have at their fingertips, not just a print tool, but also a search tool.

For this, and the other patent CD-ROM databases from RPI, we chose Dataware's 'CD-Author' software as the platform for the product. The reason for this choice is that it allows more functionality than other patent CD-ROM software available today.

What does PatentView comprise?

PatentView contains bibliographic data and images of *all* patents issued weekly by the USPTO. This includes utility, design, S.I.Rs, plant and re-issue patents. The data, both textual and graphics, are distributed on a series of compact discs. The series includes :

- the weekly issue of patents — both text and images (two discs per week, labelled A and B)

- the monthly issue of a 'Cumulating Index' — containing text only (one CD each month cumulating over the course of a year, labelled I)

- the monthly issue of 'found' images — that is images only which were previously unavailable at the USPTO (one or more CDs per month, labelled J or K).

Disc A contains all of the bibliographic data for all the patents issued in the week. It also contains the images of approximately half of those patents. Disc B contains the images of the second half of the patents and also contains the bibliographic data of these patents. Once again flexibility is the key.

This means that users with only one CD drive need only search Disc A to see what patents have been issued. If the image is needed and it is on Disc A, it can be retrieved (the user will be told). If it is on Disc B, the user is instructed to place the relevant disc into the drive. Users with two or more drives place both discs in the drives giving ready access to an offline database.

Searching over a period of time, as opposed to the latest week, is made easy due to the issue of the monthly 'Cumulating Index'.

Indexing

Each and every weekly A and B disc is fully-indexed in each and every field. Each 'Cumulating Index' disc is also indexed for convenience to the user. This disc contains the bibliographic data of *all* patents issued year-to-date. The disc is an integral part of an annual subscription to PatentView. It is the working index of PatentView and not a separate product.

Searching

There are 23 fields which can be used in combination to form either a basic search or a comprehensive search. Each field is indexed, and a 'browse' key is available.

A basic search can combine two or more fields using simple Boolean logic ('and','or' or 'not'), or by using proximity operators and truncation symbols. A comprehensive search, as good as any that can be carried out on an online host, is also available. To do this the searcher makes use of the 'F4 — connections' key, whereby the user actually enters in the search at the bottom of the search screen as opposed to filling in field labels.

Once a search has been carried out, it can be stored, rather like an online SDI but without the storage charge, and retrieved whenever new discs arrive. There is no limit to the number of SDIs a user can have.

Search *results* can also be saved i.e. exported to hard disk or a floppy disc and retrieved at a later date. The exporting of data can be done in a number of formats such as ASCII, WordStar, Lotus, Formatted Text, etc. They can also be printed directly via a laser printer.

Printing

There are numerous print options for either the text, the images, or the text and images. These include a range of documents, a range of pages from a range of documents, the image page of a current document, the first page of a range of documents, all of the documents found, etc.

What else is there to mention about PatentView?

We would like to reiterate the point about completeness. PatentView contains *all* documents issued by the USPTO. However, on some occasions, certain issued documents are not published i.e. withheld by the USPTO. When this happens, one would expect 'gaps' to appear in the CD-ROM database; not with PatentView.

When a document is not published, RPI inserts the *Official Gazette* entry into this 'gap', i.e. the bibliographic data for the document plus an abstract to search along with the exemplary claim. The image from the *Official Gazette* is also present. Quite simply it means there are *no missing patents*.

These 'found' or 'replacement' images are issued at the end of each month. A user need not worry where to look for these images because the PatentView 'Cumulating Index' disc will work transparently, directing you to the most up-to-date issue of any patent in the database, be it found, replaced, missing or simply issued regularly.

Customisation

Whilst PatentView is RPI's flagship product, it sits with our other existing CD-ROM products. *OG/Plus* is an electronic version of the *Official Gazette* of the USPTO. Again, fully searchable and indexed. *PatentHistory* is a database containing litigation details for all the patent suit cases of the most recent 17 years. Thirdly, a natural extension of PatentView has seen the development of the *PatentView Backfile*. The backfile has the full-image and bibliographic data of all patents issued since 1973. Purchased either in its complete form or as a standardised subset (as with PatentView) a user has fingertip access to up to 1,500,000 patents issued in the time period. It is the development of the backfile in an electronic format that has allowed RPI to develop along this flexible axis.

At the end of 1992, RPI produced its first truly 'customised' database, meeting the needs of one of the largest pharmaceutical companies. Merck & Co., Inc. based in Rahway, New Jersey, USA, commissioned the production of a CD-ROM database covering 28 years of US patents on five fully-indexed searchable discs, the fifth disc being a cumulating index disc updated half-yearly.

The difference between this database and any other produced by RPI, or indeed by any other information provider, is that this one comprises patents assigned to Merck and its subsidiary companies only. Produced by RPI for PInC — the Patent Information Center — the database, available on multiple copies, allows the patent information user portable, speedy and comprehensive access to a database which corresponds to approximately 50,000 pages of US patent information. The image of any of these documents is available within seconds.

RPI is currently discussing the specific needs of several companies, producing customised databases using any US or International Patent Classification, any Assignee or group of Assignees and, above all, any combination of data fields specific to the needs of a company.

Whilst this customised database contains US patent data only, RPI will endeavour to maintain this flexible approach by using other sources of patent data other than the USPTO. Looking at patents, say, from the EPO, WIPO and the UKPO and incorporating these into customised products is not alien to the developments currently under way at RPI.

Future developments

PatentScan

Developments over the last 9–12 months have recently seen the release of a new product. 'PatentScan' was launched in mid-1993 at the EPIDOS user meeting held in Brussels. PatentScan is a US patent database of all bibliographic data of *all* patents issued between 1973–92 — on one CD-ROM.

Like PatentView it is fully-indexed and searchable on a range of fields. The fields are the same as for PatentView with the exception of the claim and the abstract. For these data, RPI has created ten 'bolt-on' CD-ROMs, each containing two years of the exemplary claim and abstract for each patent.

The collection of discs allows the user to conduct a comprehensive subject search over a chosen period. Should a user require the full-image of patents found as a result of a PatentScan search, two options are presented: ordering the document via RPI's Rapid Patent service, or 'stepping-up' a subscription level to the relevant PatentView backfile year or years.

Dataware 3.1 on Windows

Other developments include the launch of Dataware version 3.1 in a Windows environment. This will herald the way forward for users to select any Windows-supported monitor and laser-printer.

MIMOSA software from Jouve Software

Perhaps one of the most important developments undertaken by RPI in the last 2–3 years will come to fruition late 1993 / early 1994. The signing of contracts with Jouve Software for the use of Jouve's new trilateral software, 'MIMOSA', means that RPI will be at the forefront of patent database technology. It is anticipated that PatentView will be available on both Dataware 3.1 and MIMOSA platforms in the new year. The latter will also allow users to search all existing ESPACE products whether current year or backfile.

Summary

By applying up-to-date technology to one of its areas of traditional strength, RPI reconfirms its position as a major supplier of patent data and documentation. Our ability to maintain a flexible approach whilst delivering complete files or custom-ised sections electronically is being continually refined.

Counter-propagation learning strategy in neural networks and its application in chemistry

Jure Zupan and Marjana Novič

National Institute of Chemistry, Ljubljana, Slovenia

Abstract

The counter-propagation learning strategy of artificial neural networks is explained. Two examples of chemical applications of the explained learning strategy are used to show how this model can yield satisfactory models for the investigated systems. In the first example a model for quantitative prediction of 'colour change' factor is developed. The second example shows the generation of the forward and inverse model for the control of a chemical process in a non-isothermic continuously stirred tank reactor (CSTR).

Introduction

In recent years, research of modelling of the complex multi-variate and multi-response systems in chemistry using the artificial neural networks has become increasingly intensive [1-5]. The general form of modelling for obtaining the sought n-value response based on the m-value query can be written in one of the following ways:

$$y_1 = f_1(x_1,x_2,...x_m, w_{11},w_{12},...w_{1p})$$
$$y_2 = f_2(x_1,x_2,...x_m, w_{21},w_{22},...w_{2r})$$
.
.
.
$$y_n = f_n(x_1,x_2,...x_m, w_{n1},w_{n2},...w_{ns}) \qquad /1/$$

or

$$(y_1,y_2,...y_n) = F(x_1,x_2,...x_m, w_1,w_2,...w_t) \qquad /2/$$

or

$$Y = \{M\} \, X \qquad /3/$$

where Y and X are multidimensional vectors, with components y_i and x_j, respectively, w_{lk} and w_k are parameters determining the functions $f_i(.)$ or $F(.)$. In the equation /3/, $\{M\}$ represents a complex non-linear operator in a form of one or more matrices composed of many parameters w_{lk}.

The above three equations, /1/ to /3/, tell us that the classical solution of this problem can be sought via a system of n non-linear equations, or *via* one or more $n \times m$ dimensional matrix(ces) and vectors composed together in various forms. Once the model is found in either of the above forms, the entire set of the system response values Y (n-dimensional response vector Y) can be calculated for **any** m-dimensional input X. The evaluation of the responses from the inputs is called the '**forward solution**' and the functions $f_i(.)$, $F(.)$ or $[M]$, the **forward model M**. In many applications, however, even more desirable than the forward model M is its inverse — the **inverse model M^{-1}** which allows to calculate X from Y.

It is generally regarded that obtaining the inverse model from the forward one is a more difficult task than obtaining the forward model. Basically, this is due to the fact that (at least from the theoretical point of view) the form of modelling functions or matrices for setting up the forward model are known to a lower or higher extent in advance while the inverse functions are much harder to express analytically.

Therefore, the methods that can provide the inverse model with the same range of errors and with the same computational efforts as they provide the forward model are extremely valuable and worthwhile to explore.

In the present work we shall try to show that besides solving the problem of forward modelling the artificial neural networks can be used for obtaining the inverse model **simultaneously** with the forward one in a very elegant and computationally efficient manner (4).

Any method, be it the classical or neural network approach, for obtaining a complex multi-variate multi-response model requires supervised learning. The name 'supervised' learning [6-8] defines a procedure in which during the training period the input **and** the requested output (response) are known in advance. Using the **difference** between the predicted output and the actual response required by the training data the system is corrected in such a way that in the next run the predicted answer (performance) of the system is better than without such a correction. The training lasts until the responses to all inputs are within the adequate tolerance limits or until the iteration quota or the number of learning attempts is exceeded. The way the mentioned difference is used for correction of internal system parameters in order to improve its performance, determines the learning strategy.

Some of the more important advantages of neural networks over the classical approaches are:

- neural networks can learn from examples, thus no explicit knowledge about models, i.e., the form of functions, is necessary

- because of their inherent non-linear response character handling of non-linear models is more 'natural' for neural networks

- neural networks can associate, i.e., they can handle incomplete or slightly corrupt data which is good for modelling outside the trained regions

- neural networks can handle continuously valued variables as well as those having discrete values

- neural networks can model inverse functions.

Until very recently, in the artificial neural network approach to the modelling problem almost exclusively the so called 'error back-propagation' [9,10] learning strategy was used. There are many different chemical applications of this method in the literature and the reader is advised to consider one of the reviews on the subject [1-5].

One of the most widely recognised disadvantages in artificial neural networks in general and in the error back-propagation learning strategy in particular is the fact that the weights in the trained neurons, representing the final synapse strengths, do not bear or have any specific meaning or relevance to the problem the artificial neural network is supposed to solve. This deficiency is not so damaging if the neural network is used as a black box — just for obtaining the sought answer. The difficulty becomes more serious if deeper, or extended, or extrapolated knowledge on the problem is pursued. In such cases the internal parameters of the model (especially if known to which actual variables are related and how) can serve to obtain the measure of quality, reliability or robustness of the entire model.

In order to overcome to some extent this deficiency we shall discuss in this paper the counter-propagation learning strategy in the artificial neural network approach to modelling.

Counter-propagation learning strategy

Because the counter-propagation [11-13,4] is one of the less commonly used neural network learning strategies, we shall describe it here more in detail. In spite of the similar name the counter-propagation learning strategy has nothing in common with the much more widely used error back-propagation method [9,10]. Due to the fact that counter-propagation is a supervised learning method, each query object is composed of two parts: input vector X and the corresponding answer or target vector Y. The neural network for counter-propagation has always two layers: the Kohonen layer to which the vector X is input and the output layer to which the target vector Y is **input**. The name 'counter-propagation' comes from the fact that the input vector X and the target vector Y are input from the **opposite** sides of the network: from the input and from the output side, respectively (Figure 1).

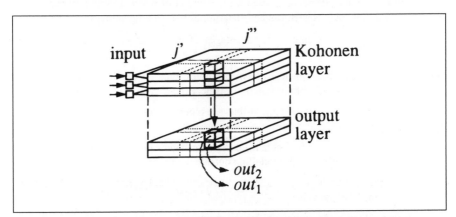

Figure 1. Counter-propagation neural network architecture. The neurons are presented as vertical columns ordered in a block. The neurons in the input (Kohonen) and in the output layer have as many weights as the input and output vectors have components, respectively. With the indices j' and j'' the position of the selected (central) neuron c is labelled

Counter-propagation network can be best visualised as two rectangular blocks composed of equal number of neurons one above the other (Figure 1). The upper block acts as a Kohonen [14,15] layer, while the second block in which the neurons have a different number of weights acts as the 'receiver' and the 'distributor' of the target information. The number of weights in neurons in the Kohonen layer is equal to the number of variables of the input vector X, while the number of weights in the output layer neurons is equal to the number of target answers.

The counter-propagation learning strategy is very simple. One step in this procedure is composed of four acts:

1 - input of object $X_s = (x_{s1}, x_{s2}, ... x_{si} ,.. x_{sm})$ to the Kohonen layer and finding the neuron c (for central) in the Kohonen layer that has the weights w^K_{ci} most similar to the input variables (equation /4/),

2 - correcting all weights of the central neuron w^K_{ci} and all weights w^K_{ji} in a certain neighbourhood of the neuron c in the Kohonen layer according to the equation /5/,

3 - using the position of the neuron c as a 'pointer' to the neuron in the output layer to which the target vector $Y_s = (y_{s1}, y_{s2}, ... y_{si} ... y_{sn})$ should be input, and

4 - correcting all weights of the central neuron w^O_{ci} and all weights w^O_{ji} in a certain neighbourhood in the output layer according to the equation /6/.

The same four acts are repeated for each input target pair $\{X_s,Y_s\}$ from the training set. Once all $\{X_s,Y_s\}$ pairs of the training set are sent through the network one epoch of learning is accomplished. The learning procedure can be repeated either until the cumulative correction of weights in the last epoch does not exceed a certain small threshold number or until the number of predefined learning epochs is accomplished.

Finding the position of the winning or central neuron c (c for central) for each vector X input into the Kohonen layer is achieved by comparing its components x_i with all weights of all j neurons in the network. The closest match defines the winning neuron c:

$$neuron\ c\ <\!\!-\!\!-\ min\left\{\sum_{i=1}^{m}(x_i - w_{ji})^2\right\} \qquad /4/$$

The equations for correcting the weights are as follows:

$$w^{K(new)}_{ji} = w^{K(old)}_{ji} + \eta(t)\ a(c\text{-}j,t)\ (x_{sj} - w^{K(old)}_{ji})$$

with j running from 1 to m $\qquad /5/$

$$w^{O(new)}_{ji} = w^{O(old)}_{ji} + \eta(t)\ a(c\text{-}j,t)\ (y_{sj} - w^{O(old)}_{ji})$$

with j running from 1 to n $\qquad /6/$

In the above equations, $\eta(t)$ is a function monotonically decreasing between two predefined values a_{max} and a_{min} with the increasing number of iteration steps t. The neighbourhood function $a(c\text{-}j,t)$ defines the percentage of actual correction (between 0 and 1) which should be applied on weights of the neuron *j* which is at the topological distance *c-j* from the central neuron *c* (Figure 2). The larger the distance *c-j*, the smaller is correction of weights. At the beginning of the learning process the neighbourhood function *a(c-j,t)* encompass the neurons and corresponding weights of the entire network, while during the learning the neighbourhood of neurons entitled to be corrected is shrinking until, at the closing epoch, the correction is applied only to the weights in the central neuron. The parameter *t* indicates the function's shrinking character.

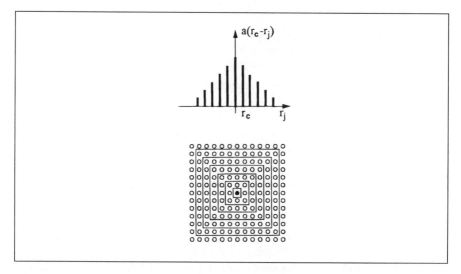

Figure 2. Square neighbourhood rings around the central (most excited) neuron is determined after each input. The number of rings in which the neurons are corrected decreases during the training.

It can be easily seen that both equations, /5/ and /6/ are virtually the same except that in the case of weights in the Kohonen neurons (superscript *K)* the correction is calculated using the components x_{sj} of the input vector X_s (equation /5/), while in the case of weights in the output neurons (superscript *O)* the corrections are calculated from the components y_{sj} of the corresponding target Y_s (equation /6/).

Examples

In order to show the ability of the method two examples will be worked out in detail. The first example is dealing with the modelling of certain quality feature of a TiO_2 based product from the measured analytical data, while the second example describes the generation of the direct (forward) and inverse model for controlling a non-isothermal continuously stirred tank reactor (*CSTR*).

The first example describes a case in which the Quality Control Department in one of our factories try to find a model that would predict the development of an undesired effect, a so called 'colour change' in their product (paint). In order to assure the quality of the product, the appearance of such undesirable property should be predicted in advance before the product leaves the factory. The data available for each sample in question were concentrations of eight most important additives and impurities that are regularly measured by the analytical laboratory in the Department.

The analytical work performed on the final product involves the analysis of eight oxide ingredients: Al_2O_3, P_2O_3, K_2O_3, Fe_2O_3, SiO_2, Sb_2O_3, Nb_2O_5, SO_3, thus, each input vector X_s describing the product s consists of eight components x_{sj} $(j=1...8)$ — concentrations of the mentioned oxides. The 'colour change' value Y_s for each product was measured relative to the other lot of the same sample not exposed to intense treatment with moisture and UV radiation. The 'colour change' value of the sample X_s, is labelled as a single component response $Y_s = (y_{s1})$. The responses Y_s were normalised between 0 and 2, 0 being the best and 2 the worst value signalling the product that changes the colour most. The value of $Y_s = 1$ is the limiting value for the quality control check. The products having the 'colour change' parameter higher than 1 are not acceptable and are either reprocessed or rejected.

Altogether complete analyses and measurements of 'colour change' of 36 products were available. From this amount 9 pairs (X_s, Y_s) were selected to make the model and the remaining 27 pairs (X_s, Y_s) were retained for test check. As it is already described elsewhere [4] the division of the entire set of objects into the training and the test set selection was made using Kohonen [14,15,4] mapping of all 36 products onto the map of 12x12 neurons (Figure 3).

By cutting the 12x12 Kohonen map into areas consisting of 4x4 neurons each the 3x3 division containing 9 areas is obtained. From each of the larger 4x4 squares only one object was selected. These are the products No.: 2, 4, 7, 13, 14, 23, 24, 27,

9	23		17 29	21	15		27		3		11 25
								33		19	13
		30									
28				36	31		32		8		1
		26					35				
34											
				24			22				4
16 18	14		12	10							7
20		6			2						5

Figure 3. Mapping of all 36 objects into the 12x12 Kohonen map. The objects are scattered quite evenly over the entire map. Only three fields contain two objects: fourth and the twelfth field in the first row and the first field in the tenth row. After the map is cut into 9 (bold lines) rectangles the selection of 9 objects representing the measurement space for building the model was arbitrary. The objects selected in the squares from upper-left to lower-right: 23, 27, 13, 28, 24, 4, 14, 2, and 7 are marked bold in double squares.

and 28. The nine selected products, each described by the 8-dimensional object vector X_s *(s=1...9)* and 1-dimensional response Y_s were then employed in the counter-propagation learning procedure using the equations /4/ to /6/. The architecture of the counter propagation network for this example is shown in Figure 4.

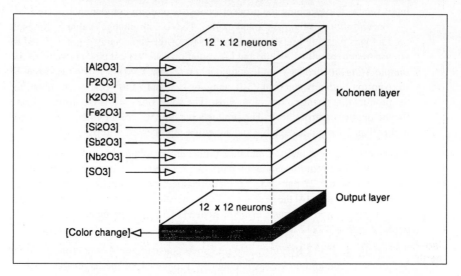

Figure 4. Architecture of the counter-propagation network for the first example. The Kohonen layer contains 144 neurons ordered in a 12x12 plane. Each neuron has 8 weights. The corresponding 144 output layer neurons have only one weight each

In this particular example we have used 300 epochs for learning 9 objects. The function $\eta(t)$ monotonically decreasing between 0.5 and 0.01 has the following form:

$$\eta(t) = (0.5 - 0.01) \frac{300 \times 9 - t}{300 \times 9 - 1} + 0.01 \qquad /7/$$

In our example, the function a(c-j,t) defines 12 rings at the beginning of the training and shrinks one ring of neighbours after each 25 epochs (25 = 300/12). The amount of weight correction depends on the neighbourhood j in which the neuron to be corrected is placed. The correction is linearly proportional to the distance c-j:, i.e. the weights of the neurons in the j-th neighbourhood are corrected with the factor $(j_{max} - j + 1)/j_{max}$. This means that the c-th neuron's weights are corrected with factor 1 $(j=1)$, while the weights of the neurons at the outer limit of the selected neighbourhood, j_{max}, are corrected with the factor $1/j_{max}$.

The resulting output layer of weights containing the 'smoothed' answers of 9 training objects is shown in Figure 5.

When the 27 test inputs (products described by eight concentration vectors) were entered into the network the following predictions were obtained (Table 1).

The correlation coefficient between the experimental and predicted values is 0.87. Due to the fact that the experimental error of measuring the 'colour change' of the

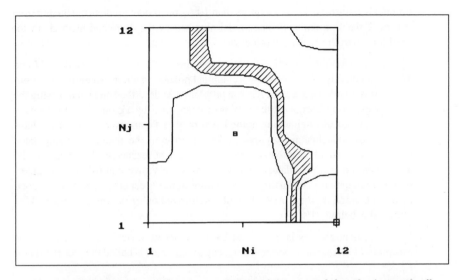

Figure 5. The resulting output layer of 144 weights containing the 'smoothed' values of nine Y_s, i.e., 'colour changes'. The lines indicate the interpolated iso-values of the 'colour changes': 0.7, 0.9, 1.1, and 1.3. The minimal value is in the centre, while the maximum is in the lower-right corner. The region between 0.9 and 1.1 iso-lines is shaded.

s	$Y_s$exper	$Y_s$pred	s	$Y_s$exper	$Y_s$pred
1	1.4	1.35	15	0.6	0.56
2	1.2	1.35	16	1.3	1.25
3	1.2	1.71	17	0.4	0.46
4	0.5	0.60	18	1.2	1.40
5	0.9	1.35	19	0.8	0.33
6	1.0	1.10	20	1.2	1.14
7	0.3	0.34	21	0.7	0.40
8	1.4	1.40	22	0.8	0.47
9	0.6	0.35	23	0.6	0.47
10	1.1	1.25	24	1.1	1.17
11	0.4	0.60	25	0.7	0.40
12	1.2	1.14	26	0.5	0.47
13	0.4	0.57	27	0.6	0.47
14	1.2	1.40			

Table I. Prediction of the property Y_s ('colour change') by the 12x12 counter-propagation network trained on 9 products

product is quite high, i.e. ± 0.2 (more than 10%), the model predicts the quantitative values of the Y_s quite well. The model obtained by the described method is now used for controlling the final products.

The second example describes building the direct and the inverse models, M and M^{-1}, respectively, for controlling a chemical process in a non-isothermal continuously stirred tank reactor $(CSTR)$. The problem was described and solved using the back-propagation learning strategy of neural networks by a group from the Department of Chemical Engineering at the University of Pennsylvania [16]. They have come to the conclusion that about $\pm 4\%$ of error in the predicted manipulated temperature is about the maximal precision that can be reached by the inverse model M^{-1} from the available data. In our work we have shown that using the counter-propagation network the quality of predictions of the direct model M can be as good as that obtained by their method, while the prediction of the inverse model M^{-1} obtained is below $\pm 1\%$.

Each chemical process is monitored by a number of variables, x_i (i=1...m). The assembly of all measured variables, x_{it}, at a given time t is called the process vector P_t:

$$P_t = (x_{1t}, x_{2t}... x_{it}... x_{mt}) \qquad\qquad /8/$$

In order to predict how the process variable x_i will vary **with time**, the variable values must be entered to the learning process as they were measured: at the consecutive time events. In terms of the 'moving window' learning technique [2,4] the 'past horizon' of the training vector, i.e. the input vector X, must contain values of, let us say variable x_i, spanning over 'previous' events $(x_{i,t}, x_{i,t-1}, x_{i,t-2}, ...)$, while in the 'future horizon', i.e. in the target vector Y, the same variable x_i should appear with its values recorded at the 'future' time intervals, i.e., $x_{i,t+1}, x_{i,t+2}, x_{i,t+3}$, etc.

In order to achieve reliable control over a given process two things are required. First, the model M must be able to predict the critical parameters of the process for a few time intervals in the future and, secondly, the corrective variable(s) must be determined from the predicted data so that the system will bounce back to the normal condition if these corrective variables are applied. The corrective or **manipulated variables** must be obtained from the **predicted** ones by the model M^{-1} which is inverse to the initial model M.

In this example a first-order reversible reaction $A \longleftrightarrow R$ in the continuously stirred tank reactor $(CSTR)$ will be studied (Figure 6).

The discussed reaction can be described by the following set of equations [16,17]:

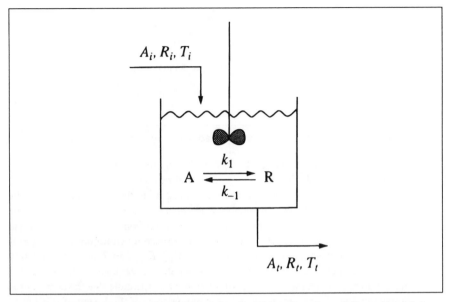

Figure 6. Non-isothermal continuously stirred reactor. A_i, and R_i are the inlet concentrations of both reactants and T_i is the corresponding temperature. A_t, R_t, and T_t are the same variables at the time t

$$\frac{dA_t}{dt} = \frac{(A_i - A_t)}{t} - k_1 A_t + k_{-1} R_t$$

$$\frac{dR_t}{dt} = \frac{(R_i - R_t)}{t} + k_1 A_t - k_{-1} R_t$$

$$\frac{dT_t}{dt} = \frac{(k_1 A_t - k_{-1} R_t)}{t} - H (T_i - T_t)$$

$$k1 = B e^{-C/T}$$

$$k_{-1} = D e^{-E/T} \qquad \qquad /9/$$

The inlet values A_i and R_i and the parameters in this system of equations /9/ have the values and units given in Table 2.

$A_i = 1\,mol/L$	$R_i = 0\,mol/L$
$B = 5 \cdot 10^3 sec^{-1}$	$C = 5033K$
$D = 1 \cdot 10^6 sec^{-1}$	$E = 7550K$
$t = 60sec$	$H = 5\,liter\,K/mol$
$T_i = 410K$	

Table 2. Values of the parameters used in the system of equation /9/

The database of state variables A_t, R_t, and T_t, required for the training and testing the model, can be calculated for any amount of time intervals by inserting the above parameters into the set of equations /9/ and integrating them. The process can be manipulated mainly by the temperature T at which the reactions run. Therefore data triplets covering many different cases can be obtained from equations /9/ by taking any given triplet of variables A_t, R_t and T_t and **randomly changing the temperature** T for the calculation of the **new** triplet A_{t+1}, R_{t+1} and T_{t+1}. This randomly changed temperature is considered as an additional variable T^m (manipulated temperature). Temperature T^m specifies the input temperature that the controller is imposing onto the system while T_t is the temperature of the system if it is left to itself.

Figure 7 shows how in the present case the process is controlled. The state variables A_t, R_t, and T_t, of the process P are measured at regular time intervals t (a). The model M checks the outcomes of the process P by regular **forward** prediction (b). For a smooth operation **no** correction is needed. However, if the controlled variable R_t (yield of the reaction) changes for some reason from the set point, the corrective action must be taken. Therefore, the **inverse** model M^{-1} calculates the corrections T^m needed (c). The correction T^m is input to the process P (d) and the actual consequences of the correction can be monitored by comparison of the process data with the set-point data and/or with those predicted by the model.

The main goal of this example is to show that the counter-propagation learning strategy of the artificial neural network trained as the forward model M can be used as the inverse model, M^{-1}, too. The counter-propagation network selected for this example is made up of 5000 neurons spread out in two (50 x 50) layers.

The size of the network was selected on the assumption that a map of 2500 Kohonen neurons would offer enough place for the accommodation of 1000 process vectors each consisting of four variables: (A_t, R_t, T_t, T^m) without resulting in too many conflicts. At the same time, the (50 x 50) counter-propagation network assures in the output layer the place for 2500 **3-dimensional** answers $(A_{t+1}, R_{t+1}, T_{t+1})$. The described counter-propagation network which has $50 \times 50 \times 4 + 50 \times 50 \times 3 = 35,000$ weights is schematically shown in Figure 8.

Selecting the central neuron c and the corrections of weights w^K_{ji} and w^O_{ji} in the Kohonen and in the output layer, respectively, was made using the equations /4/, /5/, and /6/ as described in the previous section. In the present counter-propagation training 1000 input vectors $X = (A_t, R_t, T_t, T^m)$ with 1000 targets $Y = (A_{t+1}, R_{t+1}, T_{t+1})$ were used in a 20 epochs (i.e. 20,000 inputs) long learning. The learning rate

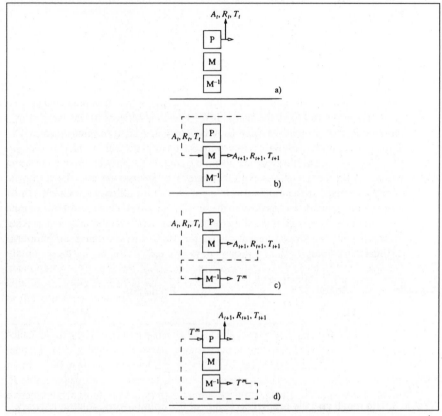

Figure 7. Control of the process *P* with the model *M* and the inverse process *M*$^{-1}$ shown in four phases

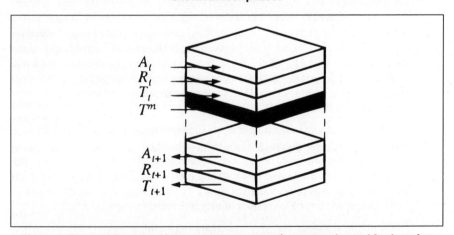

Figure 8. The architecture of the counter-propagation network used for learning the forward model *M* of the CSTR problem

$\eta(t)$ for 20 epochs and 1000 training vectors was calculated similarly to the equation /7/:

$$\eta(t) = (0.5 - 0.01) \frac{20 \times 1000 - t}{20 \times 1000 - 1} + 0.01 \qquad /10/$$

Once the counter-propagation network is trained, the retrieval of predicted process vectors is straightforward: the neuron in the Kohonen network having weights w^K_{ji} most similar to the input variables x_{ri} of the input query X_r is selected (equation /4/) as the 'winner' c. Its position c is transmitted to the output layer in which the weights w^O_{ji} of the neuron at the specified position c contain the answer. First the network was tested for the 'recall', i.e. with the retrieval of the answers after the vectors X_s which were used in the training were input as queries. Later, the trained counter-propagation network was tested with the set of thousand process vectors X_r **not** used in the training. The prediction ability for all three process variables is given in Table 3. Errors of the means are calculated for 1000 tests and have no particular applicable meaning.

Variable	Recall		Prediction	
	σ	σ^{mean}	σ	σ^{mean}
A	0.004	$1.2 \ 10^{-4}$	0.005	$1.6 \ 10^{-4}$
R	0.004	$1.2 \ 10^{-4}$	0.005	$1.6 \ 10^{-4}$
T[K]	0.86	0.027	1.03	0.032

Table 3. The recall and the prediction ability of the counter-propagation network

Let us return back to the architecture of the counter-propagation network for this study shown in Figure 8. All seven weight sheets are of exactly the same size and stacked one above the other. The **order** in which they are arranged is an arbitrary one with only one preference: the upper four sheets are used for input of variables at time t (A_t, R_t, T_t, T^m), while the lower three sheets are used for the output of the variables at time $t+1$, $(A_{t+1}, R_{t+1}, T_{t+1})$. The fact that after the training is completed, the position of weight sheets does not matter any more offers an attractive idea of rearranging them. If the weight sheets are **rearranged** in a different order, a new network is obtained (Figure 9).

The new arrangement of the weight sheets is chosen on purpose: the sheets are now divided into two new groups: the first one consisting of six weight sheets and the second one from consisting of only one. This rearranged stack of the weight sheets can be regarded as a new counter-propagation neural network having the Kohonen layer capable to accept six variables $A_t, R_t, T_t, A_{t+1}, R_{t+1}, T_{t+1}$, as the input and yield the output vector Y composed of only one variable, namely, T^m.

Without any training, just by simple **rearrangement** of weight sheets, the (50 x 50 x 6) + (50 x 50 x 1) counter-propagation network of **the inverse model M^{-1}** is obtained from the old network of the model M.

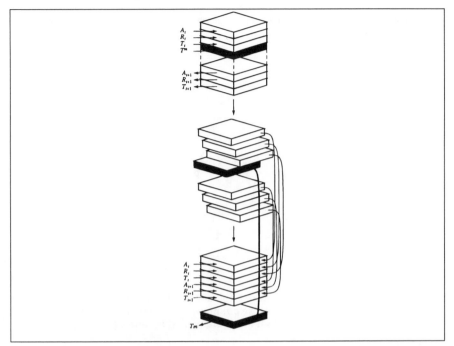

Figure 9. Rearrangement of seven weight sheets of the forward model

The rearranged network representing the inverse model M^{-1} was tested exactly in the same way as the direct model M. First by predicting 1000 T^m values from the set of data on which the original network was obtained and, second, to predict 1000 **new T^m** values from a set of newly randomly generated 1000 process vectors. The results are given in Table 4.

Variable	Recall		Prediction	
	σ	σ^{mean}	σ	σ^{mean}
T[K]	3.4	0.11	3.8	0.12

Table 4. The recall and the prediction ability of the inverse model obtained by the rearrangement of the weight sheets of the old network acting as the forward model

The prediction ability of the inverse model are worse than those of the forward model. However, they are still below 1% which means below of ±4 K.

Figure 10 shows a few random selected predictions made by the inverse model as obtained by the rearranged counter-propagation network. It is surprising to note that such a simple look-up table containing only 2500 answers which were obtained during learning the forward model gives a prediction ability of the inverse model M^{-1} within ± 1% tolerance region of the exact values. According to some authors [17] even mathematically much more sophisticated and rigorous methods for obtaining inverse models fail to produce satisfactory practical results because they use higher order derivatives and are therefore, highly sensitive to noise and numerical errors. Hence, this achievement is at least comparable if not better than most of the contemporary methods.

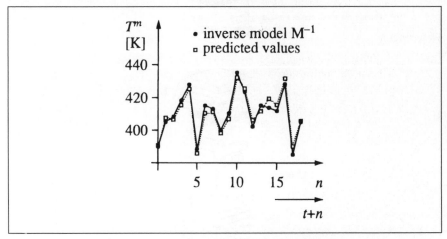

Figure 10. Comparison of test target (full circles) and predicted values (empty squares) as obtained by the inverse counter-propagation model. Targets were calculated from the system of equations /9/

Conclusions

The fact that the counter-propagation network is a look-up table, able to give merely a limited number of different answers may at the first glance appear as a serious drawback of the method. Real models yielding continuous answers in the entire range of possible values seem to be much better. However, more important than the **number** of possible answers is the **size of the error** of these answers. If the manipulated variable can vary from 380 to 420 K, the 2500 answers can cover this interval of 40 K in temperature steps that are as small as 0.016 K which is far more than the required precision of the answer. The entire problem of obtaining the correct answer is thus reduced to finding the box in the look-up table with the most appropriate answer.

In the presented method, of course, there are still a number of questions that should be answered before this method for the generation of models, and especially for generating their inverses, can generally be accepted. For example: is the retrieved prediction always the best possible one? Is it possible that in another box, showing slightly worse agreement with the input variables from that of the best, an even better answer is stored? If this is the case, how can such local minima be found? How robust is this method?, etc. After testing the method on different models and with many different applications, we hope to obtain the answer to these questions soon.

Acknowledgement

The authors are highly indebted to the Ministry of Science and Technology of Slovenia and the Bundesminister für Forschung und Technologie of the Federal Republic of Germany for financial support of this work. We wish to thank Professor

Johann Gasteiger for many helpful discussions on the counter-propagation strategy and for his kind hospitality offered to J. Z. during his stay at TUM in Garching. Thanks is due to Nineta Majcen and Karmen Rajer-Kanduc from the Quality Assurance Department of the Cinkarna Co., Celje, Slovenia who provided us with the analytical data used in the first example.

References

1. J. Zupan, J. Gasteiger, Neural Networks: A New Method for Solving Chemical Problems or Just a Passing Phase?, (A Review), *Anal. Chim. Acta,* **248**, (1991), 1-30

2. J. Gasteiger, J. Zupan, Neuronale Netze in der Chemie, *Angew. Chem.,* **105**, (1993), 510-536; Neural Networks in Chemistry, *Angew. Chem., Intl. Ed. Engl.,* **32**, (1993), 503-527

3. G. Kateman, Neural Networks in Analytical Chemistry, *Chemom. Intell. Lab. Sys.,* **19**, (1993), 135-142

4. J. Zupan, J. Gasteiger, Neural Networks for Chemists: An Introduction, VCH, Weinheim, 1993

5. P.de B.Harrington, Minimal Neural Networks: Differentiation of Classification Entropy, *Chemom. Intell. Lab. Sys.,* **19**, (1993), 142-154

6. D.L. Massart, B.G.M. Vandeginste, S.N. Deming, Y. Michotte, L. Kaufmann, 'Chemometrics: a Textbook', Elsevier, Amsterdam, 1988

7. J. Zupan, 'Algorithms for Chemists', John Wiley, Chichester, 1989

8. K. Varmuza, 'Pattern Recognition in Chemistry', Springer-Verlag, Berlin, 1980

9. D.E. Rumelhart, G.E. Hinton, R.J. Williams, 'Learning Internal Representations by Error Back-propagation', in Distributed Parallel Processing: Explorations in the Microstructures of Cognition, Eds. D.E. Rumelhart, J.L. MacClelland, Vol. 1. MIT Press, Cambridge, MA, USA, 1986, pp. 318-362

10. R.P. Lipmann, An Introduction to Computing with Neural Nets, *IEEE ASSP Magazine,* **April 1987**, 155-162

11. R. Hecht-Nielsen, Counter-propagation Networks, *Appl. Optics,* **26,** (1987), 4979-84

12. D. G. Stork, Counter-propagation Networks: Adaptive Hierarchical Networks for Near Optimal Mappings, *Synapse Connection,* **1(2),** (1988), 9-17

13. J. Dayhoff, 'Neural Network Architectures, An Introduction', Van Nostrand Reinhold, New York, 1990

14. T. Kohonen, 'Self-Organization and Associative Memory', Third Edition, Springer-Verlag, Berlin, 1989,

15. T. Kohonen, An Introduction to Neural Computing, *Neural Networks,* **1**, (1988), 3-16

16. D. C. Psichogios, L. H. Ungar, Direct and Indirect Model Based Control Using Artificial Neural Networks, *Ind. Eng. Chem. Res.*, **30,** (1991), 2564-2573

17. C. E. Economu, M. Morari, Internal Model Control. 5. Extension to Nonlinear Systems, *Ind. Eng. Chem. Process Des. Dev.*, **25,** (1986), 403-411.

Reaction type informetrics of chemical reaction databases: how 'large' is chemistry?

Bernhard Rohde

Ciba, Basle, Switzerland

Abstract. *Chemical reaction databases (CRDBs) are important tools for accessing chemical information. This paper analyses the contents of CRDBs with the methods of informetrics. The reactions in a CRDB are classified into different 'Reaction Types' (RTs), and the statistical distribution of these RTs is analysed using the REACCS Theilheimer database as an example.*

A size independent empirical probability function is presented which can be used to investigate questions such as the following:

- *How 'large' is organic chemistry?*
- *How much does the probability of obtaining a relevant answer from a CRDB query increase as the number of scanned reactions grows?*
- *What is the degree of overlap of two reaction databases from different sources?*
- *How much does user satisfaction improve by adding reaction after reaction to a CRDB?*
- *Are CRDBs large enough to enable machine learning of transform descriptions for computer assisted synthesis planning examples?*

The analysis concludes with some suggestions for the future strategies of suppliers of CRDBs with respect to contents and search mechanisms.

Introduction

The term informetrics covers "the study of quantitative aspects of information in any form" [1] . Using informetrics, we can, for example, look at the spelling of just this term [2]. A review of the frequencies of the two words (or related terms) in the database indices of STN/Karlsruhe yields Table 1.

According to these data, informetrics appears to be the more common spelling, especially in the literature on information science and information work covered by the database INFODATA.

This paper applies informetric methods to chemical reaction databases using a classification method called the 'Reaction Type.' Similar methods are discussed in [3].

A short sketch of the history of the study is given after this introduction. Then, we will illustrate the principles guiding the reaction type definition used in this study, discussing some examples. Some basic statistics will be presented followed by the introduction of the Reaction Type Probability Function. The discussion focuses on the Theilheimer database sold by Molecular Design Ltd. and selected other data-

Database	INFOmetr?	INFORmetr?
Inspec	3	18
CA	2	0
Compuscience	0	20
BLLDB	0	2
CONFSCI	0	2
CONF	1	3
DISSABS	0	6
Infodata	1	83
Total	**7**	**134**

Table 1: Different spellings of Informetrics and derived terms in the keyword indices of various databases at STN

bases when this is necessary to illustrate a specific point. The next section will discuss how the virtual 'size of chemistry' could be defined using this function and how the degree of overlap of two databases can be defined. The informetric results are then used to derive relations between the database size/cost and its utility. They will also shed some light on the use of CRDBs for machine learning of transforms for computer assisted synthesis planning. The final section will give some suggestions on abstraction policy and search methods for database vendors.

The space available here does not allow a full description of all the details involved in reaction type definition, coding, and canonization, the derivation of statistical formulae, numerical mathematics, and programming, and the full data on all databases analysed. For a more complete description of this work, refer to the series of three papers [4], which are in preparation.

The history of the study

The roots of this work are in a synthesis planning project conducted by a consortium of German and Swiss chemical and pharmaceutical companies, namely: BASF, Bayer, Ciba-Geigy, Hoechst, Merck, Roche and Sandoz. These companies developed a program called CASP for Computer Aided Synthesis Planning, that was derived from the SECS program by T.Wipke [5].

In the early 1980s, synthesis planning programs such as CASP were major sources of reaction information, which was stored as so-called 'transforms' and written in a chemical programming language. Chemists presented the program with the structure of their synthetic target plus some 'strategic bonds' which were allowed to change during the desired transformations. In a one step synthetic analysis, CASP then returns the transformed structures for those transforms in its library which would fit structure and strategic bonds. The reference information stored with the transforms then pointed to the primary literature.

During the late 1980s, various vendors developed chemical reaction databases (CRDBs) together with sophisticated search software [6], which challenged the similar use of synthesis planning. Therefore, the CASP pool started an initiative to connect synthesis planning with reaction retrieval [7]. One of the projects associated with this initiative tried to develop a classification scheme for reactions which was to be as similar as possible to the criteria used to assign individual reactions to transforms. The result of this project was the definition of the Reaction Type that is described in this paper.

The immediate goal of this classification was to facilitate the correlation of reactions from CRDBs with CASP transforms, since reactions sharing a reaction type are likely to be correlated to the same transform [8]. In addition, the resulting correlation together with uncorrelated reactions of the same reaction type could be used to improve the transform library systematically because they indicated deficiencies in the transforms.

In the distant future, the reaction type could be used as the basis for a machine learning program. It would infer transforms or some other representation of generalised reactions from examples in CRDBs, which is already being studied at a number of places [9]. Such a procedure is necessary because it cannot be commercially justified to develop a synthesis planning program from scratch using manually deduced transforms. The current CASP transform library written in the ALCHEM language comprises more than 50 man-years of work, which would be lost if ALCHEM were not supported by any future program. However, ALCHEM is a programming language, which, as such, can only be executed in one, the retro-synthetic, direction. It is complex, not expressed in graphical terms of chemical structures, and difficult to debug. Any reliable inference in synthetic direction, like side reactions or multiple active positions in the reactants, would require extension of the language and consequently manual (and costly) updating of transforms.

Machine learning from experimental facts can only be as good as the collection of examples used. In the case of automatic transform writing, the number of examples must be large enough to represent the class of chemistry the transform represents. Statistical information on the number of examples for each reaction type is therefore valuable for predicting the number of transforms that can be supported by enough examples from the reactions in one or more reaction databases. This paper uses informetric methods to make such predictions.

What is a Reaction Type?

The Reaction Type (RT) of a reaction represents the significant characteristics of a chemical reaction. It is denoted in a language similar to a reaction substructure query and contains the reacting atoms and bonds augmented by the chemically important neighbourhood. It also includes ring formation and ring cleavage information, which mimics the synthesis planning view of CASP.

Stereochemistry has been ignored in this study for two reasons: 1) The CASP transform library is very poor with respect to stereochemistry, and therefore it would not help in the combination of REACCS with CASP. 2) The quality of the stereochemical description of the reactions in the REACCS databases as well as the number of stereochemically described RTs is not sufficient to justify the effort. However, the coding method for RTs, called the GM-Description [10], is capable of encoding, canonizing and hashing stereochemical information.

This project did not attempt to invent a new method to map reactant atoms to product atoms of a reaction description, but relies on the mapping provided by the reaction databases, namely the REACCS databases. This information has largely been generated automatically and is incorrect for some reactions, but the overall informetric conclusions are not affected, because they deal with the numbers and not the precise formulations of the RTs.

The RT of this study is defined in a one step hierarchy. The so-called MiniRT comprises the reacting atoms and bonds, i.e. the 'reaction centre.' Some of the atoms are generalised to generic atom types, e.g. the halogens are mapped to an 'X' atom type. The higher level FullRT uses specific atom types for the reaction centre and adds pieces of the environment according to a set of rules. Those rules are described in detail in [4].

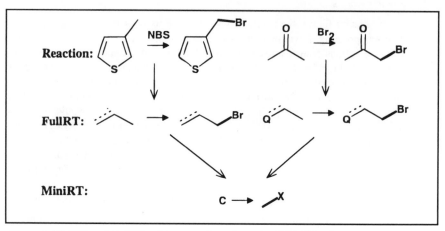

Figure1: MiniRT and FullRTs of two reactions from the ORGSYN database of REACCS. Reacting bonds are drawn with thick lines.

The two reactions in Figure 1 illustrate the hierarchical definition: both reactions are substitutions of hydrogen at a carbon atom, which is expressed by the common MiniRT. The activation of the reaction centre is covered by the environment included in the FullRTs. They are different for the two examples, because the rules defining the FullRT consider the carbon-carbon double bond of the thiophene ring to be different from the carbon-oxygen bond of the carbonyl group. The bond order of the two double bonds is generalised to 'multiple,' which is denoted by a dotted second line. Carbon-carbon single bonds which do not contribute to the activation or deactivation are not made part of the FullRT except for special cases such as the distinction between reactions at the carbonyl groups of ketones and aldehydes.

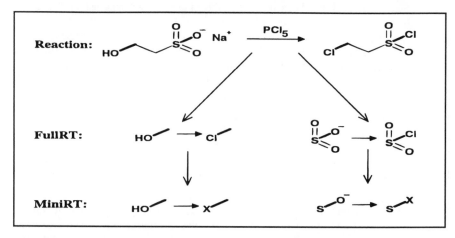

Figure 2: One reaction containing two independent RTs. Example drawn from ORGSYN database.

One reaction can also give rise to more than one RT. Figure 2 shows a reaction replacing oxygen by chlorine at two independent sites of the molecule. Other examples of multiple RTs include simultaneous reduction of multiple aromatic nitro groups or deprotection of functional-groups during work-up.

The RTs of a reaction database are also divided into four classes, which are similar to the classes of chemistry covered by the different sections of the CASP transform library. The classes are:

CC_MAKE   for carbon-carbon bond making reactions,
CC_BREAK   for carbon-carbon bond breaking reactions,
QC_MAKE   for carbon-hetero bond making reactions, and
FGIA   for functional-group chemistry.

They relate to the 'synthetic value' of the transform or RT, e.g. CC_MAKE reactions are building the carbon skeleton, while FGIA reactions just modify the functional groups. These reaction classes form a hierarchy, i.e. if one class applies, the classes listed below cannot also apply. Figure 3 shows one example from the Theilheimer database for each of these classes.

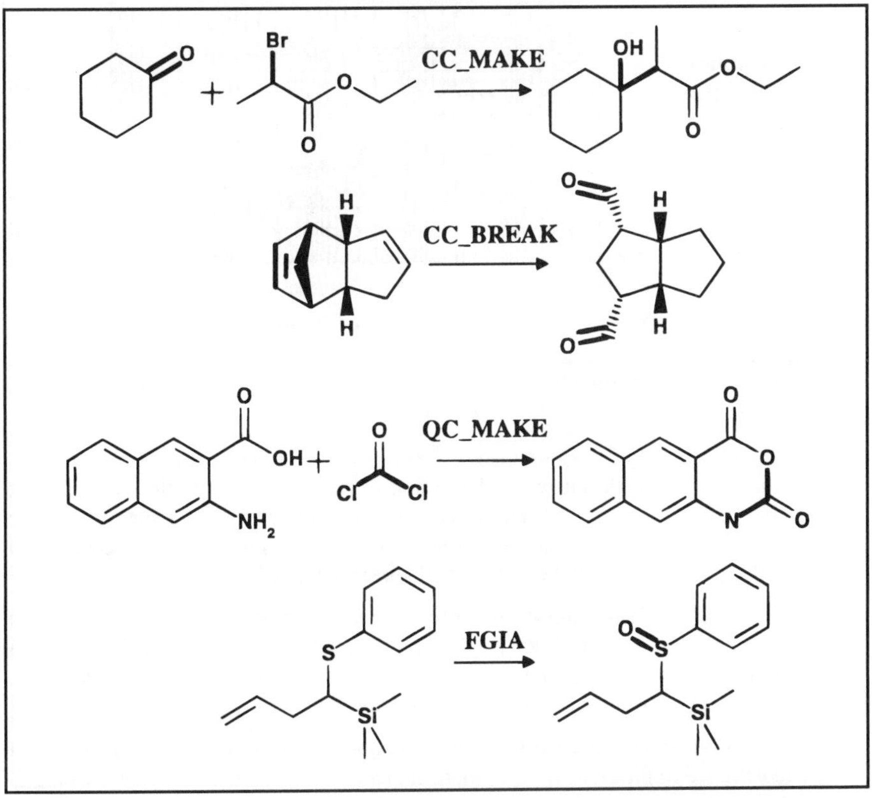

Figure 3: Examples for the reaction classes drawn from the Theilheimer REACCS database.

Some basic statistics

The reactions of some of the REACCS databases have been analysed using the RT described above. Reactions with more than one product, dubious mapping, or other problems have been excluded from processing. This section presents some statistical results that can be derived by counting various properties of the RTs of a reaction database. Some of these empirical distributions form the basis of the theory of the Reaction Type Probability Function developed in the next section.

If the reactions of a database are classified by their RT, one may count how many reactions are examples of a given RT [11]. Then, the RTs can be ordered by decreasing example counts and the number of examples of an RT can be plotted against its ordinal number. Figure 4 shows one graph for the classification by FullRT and one for the MiniRT.

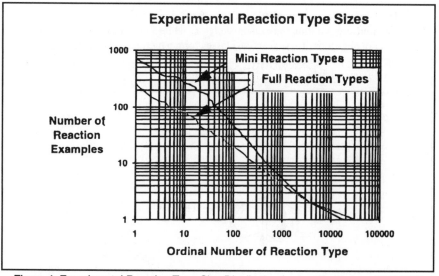

Figure 4: Experimental Reaction Type Size Distribution for the Theilheimer REACCS database.

There, for example, approximately 700 reaction examples of the most frequent MiniRT, which is the cleavage of a carbon-oxygen bond leaving the oxygen as the only cited product, most often the saponification of an ester of a simple acid. On the other hand, there are somewhat more than 30,000 FullRTs (represented by at least one reaction example) in the Theilheimer database.

If RTs are considered significant only when they can be reproduced in more than one reaction, then some 4,000 FullRTs have been proven by the Theilheimer to be significant. This also means that there are about 26,000 FullRT singletons, i.e. curiosities with only a single reaction example.

Another interesting figure is the average number of FullRTs per MiniRT. If all RTs are included in the analysis, then there are less than two FullRTs for each MiniRT, but as we have seen, RT counts are dominated by singletons. The average rises to

3.5–4 FullRTs per MiniRT, if MiniRT singletons, which are also FullRT singletons, are neglected (see Table 2).

	FGIA	CC_MAKE	QC-MAKE	CC_BREAK	ALL
All RTs	1.94	1.68	1.75	1.64	1.80
Without singletons	3.57	4.00	3.81	3.64	3.73

Table 2: Average number of FullRTs per MiniRT listed by reaction class.

Figure 5 shows the distribution of the four reaction classes in the Theilheimer database. The reactions of the largest class are functional-group interconversions. This fact becomes even more pronounced when singleton RTs are excluded, because the constructive classes CC_MAKE and QC_MAKE have a relatively larger portion of 'curiosities.'

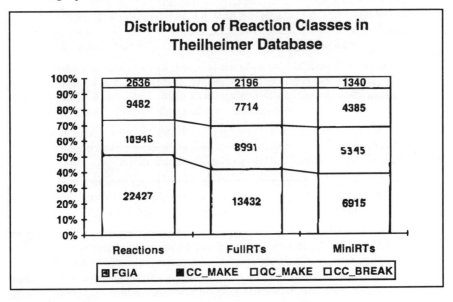

Figure 5: Relative numbers of reactions, FullRTs, and MiniRTs for the four reaction classes.

The Reaction Type probability function

The previous section discussed size-dependent measures of a reaction database. For predictions, we need size-independent characteristics of the abstraction policy of a database vendor. The 'Reaction Type Probability Function' (RTPF) is such a measure. It associates with each conceivable RT i its probability $p(i)$ that a random reaction, selected from literature to be included in the data collection under consideration, will have exactly this reaction type. Without loss of generality, we can assume that the RTs are indexed in order of decreasing probability, i.e. $p(i) > p(i+1)$.

If we assume that $p(i)$ can be approximated by some function $f(i,a,b,c,d,...)$ with free parameters a, b, c, d, ..., then we can estimate the free parameters using the frequency data computed by RT analysis of the CRDBs. In general, the free parameters will be different for different databases, because they encode the abstraction policy of a given database vendor.

During this study, a number of different RTPF approximations have been tested. We will only show two of them, including the currently best functional form. If there were a model describing which RT will be reported in the literature and why it is selected for a database, then we could use and fit the resulting functional form. Currently, we can use standard statistical probability functions, which perform poorly, and which have not been derived for probabilities as a function of an ordinal number, or construct some function which fits the data well.

When fitting the free parameters to the data, we can use both ends of the empirical frequency function (see Figure 4). The higher, most frequent end is used under the assumption that the order of the RTs in the RTPF matches the order in the empirical frequency function and, therefore, each frequency is compared to the correct probability. This is certainly not true for less frequent RTs, and only as an approximation for the others. Therefore, we use only the first 20 RT frequencies and with a smaller weight than for the counts of the tail of the frequency distribution.

The tail can be modelled more rigorously. Given an RTPF $p(i)$ and a database size n, we can derive expressions for the number of RTs having at least one example $F_0(n)$ and those having exactly k examples $F_k(n)$ given in Figure 6. These quantities can be directly compared to the corresponding experimental counts.

The model of this theory is a chemistry which is static in its principles. It would not hold any more when new breakthroughs such as the development of stereoselective synthesis or the invention of boron chemistry would appear.

Another problem may occur when the conventions for structure representation change over time because it would also change the representation of some RTs which are in fact identical.

In this study, we analyse the database vendor's view of chemistry, which is only an approximation of chemistry in general and which might change over time. This subjective view will probably magnify some parts and reduce others. However, most of the conclusions presented here are either qualitative in nature or involve the comparison of databases, where just this different view is the subject of discussion.

$$F_0(n) = \sum_{i=1}^{\infty} (1 - (1 - p(i))^n)$$

$$F_k(n) = \sum_{i=1}^{\infty} \binom{n}{k} p^k(i)(1 - p(i))^{n-k}$$

Figure 6: Expressions for the number of RTs with at least one example $F_0(n)$ and exactly k examples $F_k(n)$ in a database of n reactions

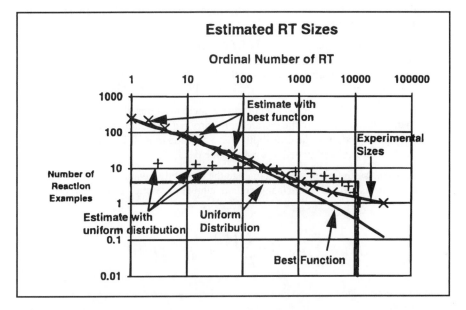

Figure 7: Comparison of the quality of fit for two approximations of the RTPF. The FullRT statistics for the Theilheimer database are used for fit and simulation

The most objective sample of organic chemistry in general would be the CAS reaction database, which is collected for documentation purposes and not with a special audience in mind, but we did not have access to these data. The other extreme, an automatically pre-processed database, would be ChemReact from InfoChem which was pruned to one example per RT using a definition similar to the one described in this paper.

Figure 7 illustrates the results of fitting two different functional forms. The experimental size distribution of the FullRTs in the Theilheimer database is copied from Figure 4. The first functional form is the uniform distribution of 11,200 RTs, i.e. all RTs are assumed equally likely and their number is 11200. The chart shows the uniform distribution multiplied by the total number of reaction examples and selected data points (+) of a simulation for this distribution using random numbers.

The currently best functional form, which is given in Figure 8, is also displayed as a scaled probability function and as a sequence of data points (x) selected from the simulation. Note that although the simulation results correspond quite well with the experimental data, only the head of the function is close to the experimental frequency distribution. The tail of the probability function represents all conceivable

$$p(i) = \frac{A}{e^{ai^{1/b}} + c}$$

Figure 8: Currently best functional form for fitting the RTPF. A is a normalisation factor and a, b, and c are the free parameters.

RTs while the tail of the frequency function contains only RTs represented in the Theilheimer database.

The next assumption we will make for the following discussion is that a researcher querying a database looking for an RT will do this with the same probability as the input clerk would input a reaction of this RT. The rationale behind this assumption is as follows:

If the reaction database just documents literature reactions, if the database does not change the intentions of the researcher to carry out a reaction, and if the reaction is to be published, then the probability that a reaction or its RT is searched for is the same as the probability that it is put into the database.

With this assumption, we can compute the probability that a random query posed to a database will hit more than a certain number of reaction examples as shown in Figure 9.

$$p_{>k}^{n} \doteq \sum_{i=1}^{\infty} p(i)(1 - \sum_{l=0}^{k} \binom{n}{l} p^{l}(i)(1 - p(i))^{n-l})$$

Figure 9: Probability that a random query to a database of n reactions will yield more than k hits.

A certain number of reactions of a CRDB are singletons, i.e. reactions whose RT is not shared by any other reaction in the database. Figure 10 depicts the probability that a reaction being added to database defines a new RT as a function of the number of reactions already present. We can see that for the size of the Theilheimer database (~45,000 reactions) there is a 30% chance that a new reaction defines a new bond change (MiniRT) and 60 % to open a new FullRT. For the CASREACT database, the figure predicts that 35% of the new reactions have novel FullRTs. The observation of InfoChem, made while they created the ChemReact database from a collection of 1.8 million reactions, that 20-30% of the added reactions create a new RT also fits into this prediction.

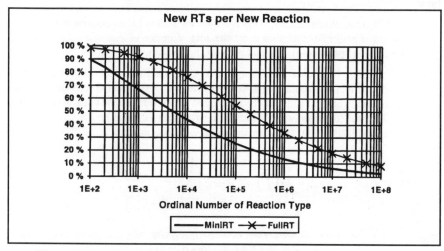

Figure 10: Percentage of novel RTs when reactions are added to a database

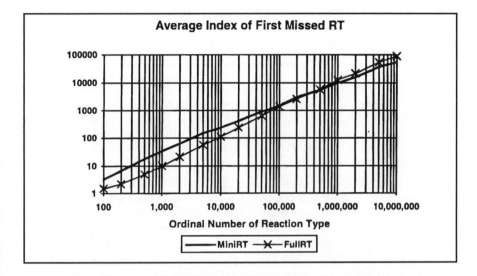

Figure 11: Average index of the first RT not supported by at least one example

In the RTPF, the reaction types are ordered by increasing probability, which could also be interpreted as increasing importance. Figure 11 illustrates how likely it is to miss an important RT. It plots the average index of the first RT not included in the database as a function of database size. This relation can be found by solving the equation of Figure 12 for the index k.

From the graph of Figure 11, we see that for purely statistical reasons on the average one out of the most frequent 1000 MiniRTs (700 FullRTs) will not be present in a database of ~50,000 reactions like the Theilheimer. To reduce this to one out of 10,000, we would require a million reaction records.

The virtual size of chemistry

The RTPF is a database size-independent characteristic of the reaction database. It is possible to derive other such characteristics from the RTPF.

An interesting number would be 'the' size of the chemistry covered by the reaction database. If all RTs were equally likely, i.e. if the RTPF was the uniform distribution, then the size would be the defining size parameter of this approximation function. However, as we have seen before, the uniform model does not fit the data well. On the other hand, the functional forms that explain the experimental counts well are all infinite, i.e. they assume infinitely many conceivable RTs.

$$\min_{k}\left(\left|1 - \sum_{i=1}^{k}(1-p(i))^{n}\right|\right)$$

Figure 12: Defining expression for the average index k of the first RT not supported by an example reaction in a database of size n

$$G_1 = \sum_{i=1}^{\infty} p^2(i) = \frac{1}{m}$$

Figure 13: Size of chemistry computed from a one reaction database

Despite these observations, we can still define numbers which could be called 'sizes'. They will measure different aspects of the database. They only match for all definitions in the case of the uniform distribution. They may even differ by orders of magnitude between definitions, but may be used to compare two databases when the definition is the same. This is similar to the difference between the diameter and length dimensions of a tube.

The idea behind the derivation of these sizes is as follows: Let m be the desired size. Then we derive a formula supposing the uniform distribution with parameter m for some characteristic numeric property G of the collection of RTs. Given this formula, we can then compute the numerical value of the property from the best approximation of the RTPF instead of the uniform distribution and solve the formula for m. We shall expand on two such properties here.

The first one is the probability of hitting the RT of the reaction example of a database containing a single reaction when adding a new reaction. The general expression for this probability, as a special case of Figure 9, and the result for the uniform distribution is given in Figure 13.

For the second property, we derive an expression for the hit rate for a database with m reactions. From Figure 9 we can conclude that this fraction is $(1-e^{-1})$ for large m. Therefore, to derive the value of G_2 for some probability distribution, we systematically vary the size of the database in the formula of Figure 9 until the hit rate approximates $(1-e^{-1})$. Figure 14 summarises the definition of G_2.

$$G_2 = m = n,$$

$$\text{with } n \quad \text{for which}$$

$$1 - e^{-1} = \sum_{i=1}^{\infty} p(i)(1-(1-p(i)^n)$$

Figure 14: Size of chemistry computed from database size
with hit rate $(1-e^{-1})$

	From G^1		From G^2	
	FullRT	**MiniRT**	**FullRT**	**MiniRT**
CLF	6268	633	346,411	22,227
THEIL	6236	692	698,306	22,248

Table 2: Estimates for the size of the chemistry covered by the Current Literature File
and the Theilheimer

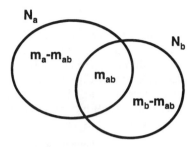

Figure 15: Database overlap parameters

The *m* values derived from the two properties for the Current Literature File (CLF) and the Theilheimer (THEIL) database are shown in Table 2.

G_1 measures the size of chemistry with an emphasis on the most frequent RTs, while G_2 gives a more realistic estimate. The CLF is stronger than the Theilheimer in the most frequent RTs, which reflects the fact that the Theilheimer selection process intentionally reduced the number of reactions for common RTs to a representative collection. The results for G_2 show that the Theilheimer covers more chemistry in general. The MiniRTs, which mirror only the changing bonds, show no significant difference. We can conclude that the Theilheimer will show more chemical variation for the same bond change.

The size measures derived using G_2 can be used to estimate the overlap of the chemistry covered by two databases. Figure 15 illustrates the parameters used to describe overlap. The assumption is that two databases a and b are mixed, that they have uniform RTPF with parameters m_a and m_b respectively, and that their sizes are N_a and N_b. They overlap in a region covering m_{ab} RTs. If we derive the size m from the G_2 property of the combined database, then we can estimate the database overlap m_{ab} by solving the equation of Figure 16 for m_{ab}.

For the CLF and Theilheimer databases, we have $m_a = 346,411$ ($N_a = 30,702$) and $m_b = 698,306$ ($N_b = 45,491$), respectively. The combined statistics yields $m = 720,543$ from which we conclude that $m_{ab} = 238,647$. In other words, CLF includes 34% of the Theilheimer chemistry while the Theilheimer covers 69% of the CLF RTs. If overlap was not measured by overlapping RTs but weighted by their probability of being hit by a random query, then a larger overlap value is likely.

$$m(1-e^{-1}) = (m_a - m_{ab})\left(1-\left(1-\frac{N_a}{N_a+N_b}\frac{1}{m_a}\right)^m\right)+$$

$$(m_b - m_{ab})\left(1-\left(1-\frac{N_b}{N_a+N_b}\frac{1}{m_b}\right)^m\right)+$$

$$m_{ab}\left(1-\left(1-\frac{N_a}{N_a+N_b}\frac{1}{m_a}-\frac{N_b}{N_a+N_b}\frac{1}{m_b}\right)^m\right)$$

Figure 16: Defining equation for overlap parameter m_{ab} using property G_2

Database cost versus utility

Figure 17 plots the hit probability for MiniRT and FullRT type queries against the size of the database using the equation of Figure 9 and the parameters fitted to the Theilheimer database. For the more general MiniRT query, we can see that current in-house reaction databases, which collectively contain approximately 300,000 reaction, yield a hit rate ~65% if a single hit will satisfy the user and ~40% if ten hits are necessary. Only 35% and 10% hit rate can be expected for the more specific FullRT query requesting one or ten hits, respectively.

The graphs of Figure 18 show how the increase in probability to satisfy a MiniRT or FullRT (x) query per 1000 new reactions relates to the number of already present reactions. This relation can be translated into financial terms using a conservative estimate of 30.- sFr per record for reaction input. Figure 18 shows this as thin lines.

The current CASREACT database contains approximately 1.1 million reactions with a growth rate of 50,000 to 100,000 reactions per year. Using Figure 17 and Figure 18, we can estimate that more than 85% of the MiniRT (FullRT 65%) can be satisfied and this performance increases by 0.3% points (FullRT: 1% point) per year.

One goal of this study was to investigate the feasibility of machine learning of generalised reactions (transforms) for synthesis planning. The analysis of the RTPF can answer the question: How large must a database be to provide at least a sufficient number of examples. Figure 19, which can be derived from the equations of Figure 6, plots the number of RTs with at least 5 and at least 20 examples for both the FullRT and the MiniRT case.

The FullRT was designed to mimic the CASP transform. If we require at least 20 reaction examples to explore the variability of a transformation, the Theilheimer database can provide data for 100 transforms. To be able to infer 1000 transforms would require a database of approximately 300,000 records. The CASP transform library contained some 6000 transforms, which, however, sometimes share one RT. More than 1 million reactions would be required to learn that number of transforms from examples.

Suggestions

Figure 17 and Figure 18 illustrate the diminishing returns of reaction database production. When the database is only searched by RSS which is close to an RT search, the user satisfaction can be increased above current levels only with great effort.

Since the pure input of reaction records will not be cost effective, we propose three ways of approaching the problem:

- The database vendor could decide to produce small specialised databases which distinguish themselves from the large collections by their contents of non-structural data. There might be generalisation hints, information on the common use of the reactions, or cross-references between different records.

  Since the database is specialised, the user will only expect to receive a hit when the database is likely to contain it. Those hits are of higher value than the usual literature references because of the high-quality data associated with them. The Protecting Groups Database by Synopsys is one such example.

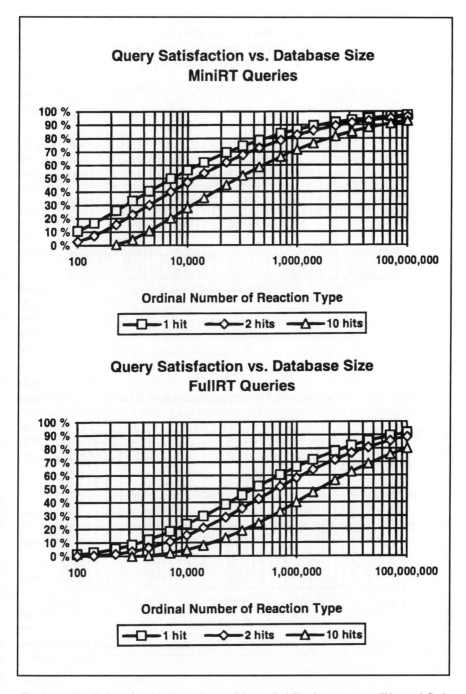

Figure 17: Predictions for the dependence of the probability that a query will be satisfied as a function of the size of the reaction database. The MiniRT and FullRT are model generic and specific queries, respectively. The different lines represent queries looking for at least 1, 2, or 10 answers

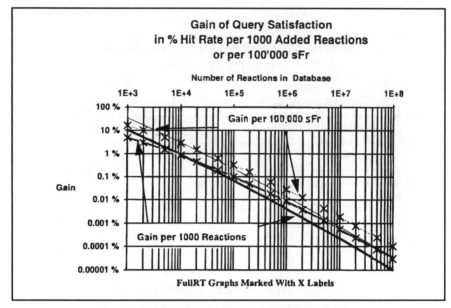

Figure 18: Gain in hit rate per 1000 added reactions or
100'000 sFr assuming 30 sFr per Reaction

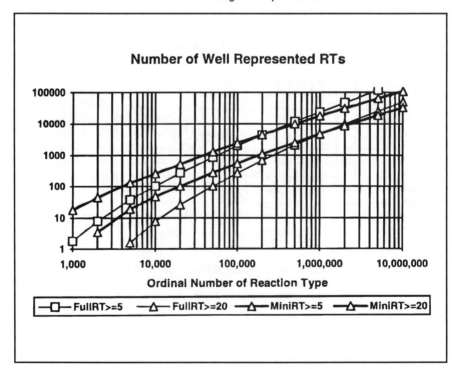

Figure 19: Number of RTs with at least 5 and at least 20 examples for MiniRT and FullRT

- Secondly, the reaction data for the database could originate from a source which collected it for a different reason such as documentation and, therefore, has already paid for it. The CASREACT database is an example because the structures have already been input for the registry system. The Beilstein database also links reactants and products and could be transformed into a source of reaction records.

- Finally, we want to suggest a novel search method based on RT analysis. Figure 4 shows that a large portion of the database contains rare RTs which cannot be exploited when looking for a single RT. Reaction database vendors have been using similarity searching to overcome this problem. We propose another method which will increase the use of already-abstracted reaction records.

The method might be called 'Reaction Type Substructure Search' (RTSS) and resembles a one step synthesis planning. In some sense, it is the inverse of a Reaction Substructure Search. The user inputs the synthetic target molecule. The algorithm, then, searches for those reactions which have an RT whose product-substructure is part of the target molecule. The matching parts of the molecule are checked to determine whether their environment would yield exactly this product RT if RT analysis were applied to it. If this is true, the reaction is considered a hit.

One problem with this approach is that some RT products will match in most targets and that, unfortunately, those are the most frequent ones. Therefore, it is necessary to sort the resulting hits to show the most interesting reactions first. The program might use the 'synthetic value' implied by the reaction class and a functional-group based similarity measure for this purpose.

With this approach, all parts of the target can be simultaneously used as queries for the search and more RTs will contribute to the answer set.

Conclusion

This paper has shown that it is possible, under simplifying assumptions, to derive a database size independent characteristic function of the contents of the database. This function was used to define quantities like different 'sizes' of chemistry. It has been applied to predictions about computer assisted synthesis planning, and it was shown helpful for the business of database vendors. Finally, a novel method of searching reaction databases has been sketched which could become 'Poor Man's Synthesis Planning.'

Acknowledgement

I want to thank R. Wehrli at Ciba for his support especially at the beginning of this project, the CASP consortium for adopting my work as part of a project enabling me to continue, C. Chylewski, N. DeMesmaeker, K. Kocsis, and D. Schulz for their contribution in refining the idea of the reaction type, and Ciba for the possibility to finish the study although Ciba left the CASP consortium by the end of 1990. Finally, I want to thank Molecular Design Ltd. for allowing me to process the contents of their databases.

References and notes

[1] J. Tague-Sutcliff, Special Issue on Informetrics, Editorial Note, *Information Processing & Management* **28**, 1-3 (1992).

[2] W. A. Turner, 'Infometrics for mapping and measuring science and technology,' pp. 113 - 131 in *Proceedings of the Montreux International Chemical Information Conference & Exhibition*, H. Collier, Ed., 1991, Springer Verlag: Heidelberg, uses one spelling while this paper uses the other.

[3] a) S. Fujita, 'Structure-Reaction Type Paradigm in the Conventional Methods of Describing Organic Reactions and the Concept of Imaginary Transition Structures Overcoming this Paradigm,' *J. Chem. Inf. Comput. Sci.* **27** (3), 120-6 (1987).

   b) E. S. Blurock, 'Computer-Aided Synthesis Design at RISC-Linz: Automatic Extraction and Use of Reaction Classes,' *J. Chem. Inf. Comput. Sci.*, **30** (4), 505-10 (1990)

[4] a) C. Chylewsky, K. Kocsis, N. de Mesmaeker, B. Rohde, R. Wehrli, D. Schulz, 'Reaction Type Informetrics. 1. Definition of Reaction Types: A Structural Classification of Reactions,' in preparation.

   b) B. Rohde, 'Reaction Type Informetrics. 2. Canonization of Reaction Types: An Automatic Reaction Classification,' in preparation.

   c) B. Rohde, 'Reaction Type Informetrics. 3. Statistical Results: The Virtual Size of Chemistry,' in preparation.

[5] W. T. Wipke, H. Braun, G. Smith, F. Choplin, and W. Sieber, 'SECS - Simulation and Evaluation of Chemical Synthesis: Strategy and Planning,' pp. 97 - 127, in W. T. Wipke, W. J. Howe (Eds.), *Computer-Assisted Organic Synthesis*, ACS Symposium Series No. 61, Washington (1977).

[6] a) J. H. Borkent, F. Oukes, J. H. Noordik, 'Chemical Reaction Searching Compared in REACCS, SYNLIB, and ORAC,' *J. Chem. Inf. Comput. Sci.*, **28** (3), 148-50 (1988).

   b) E. Zass, S. Mueller, 'New Possibilities for Research on Organic Chemical Reactions: Comparison of In-House Databank Systems, REACCS, SYNLIB, and ORAC,' *Chimia*, **40** (2), 38-50 (1986).

   c) P. Bador, M. N. Surrel, 'Computer Systems for Searching of Chemical Reaction Databases and Systems for Computer-Aided Design of Organic Synthesis,' *New J. Chem.*, **16** (3), 413-23 (1992).

[7] J. B. Hendrickson, T. M. Miller, 'Reaction Classification and Retrieval. A Linkage Between Synthesis Generation and Reaction Databases,' *J. Am. Chem. Soc.*, **113** (3), 902-10 (1991).

[8] A reaction is said to correlate with a transform when CASP would propose the reaction using that transform. This correlation should be used to support the suggestions of CASP using examples from CRDBs while the chemist browsed the CASP results.

[9] (a) G. Kaufmann, P. Jauffret, C. Tonnelier, T. Hanser, 'Development of Computer Tools for Machine Learning of Generic Reactions Starting with Specific Reactions,' *Chem. Inf.*, **2**, 1-11 (1990).

(b) H. Gelernter, J. R. Royce, C. Chen, 'Building and Refining a Knowledge Base for Synthetic Organic Chemistry via the Methodology of Inductive and Deductive Machine Learning,' *J. Chem. Inf. Comput. Sci.*, **30** (4), 492-504 (1990).

(c) E. S. Blurock, 'Automatic Extraction of Reaction Information from Databases Using Classification and Learning Techniques,' *Chem. Inf.*, **2**, 25-35 (1990).

[10] B. Rohde, 'GM-Search: A System for Stereochemical Substructure Search,' Dissertation, University of Zürich 1988.

[11] The RT descriptions coded in an RSS query language are converted to GM-descriptions [10], canonized, and hashed. The comparison of RTs is done only using these 48 bit hash codes, which is precise enough for the present purpose.

Learning about synthetic methods in organic chemistry: the GRAMS project

Ph. Jauffret, Th. Hanser, J.F. Marchaland and G. Kaufmann

Laboratoire de Modèles Informatiques Appliqués à la Synthèse (URA 405 du CNRS), Université Louis Pasteur 4, rue Blaise Pascal, 67000 Strasbourg, France

Introduction: learning in Computer Aided Design of organic synthesis

In organic chemistry, molecules are obtained from available starting products by a sequence of chemical reactions called a synthetic pathway. Searching for a new pathway is complicated because of the great number of possible starting materials and already known synthetic reactions, and because inferring general rules is difficult. Due to the experimental nature of chemistry, it is generally difficult deciding on the feasibility of a reaction.

Before beginning the experimental work, the chemist draws one or several theoretical pathways. To help him at this stage of planning, some programs have been developed by different researchers in a new scientific area: the Computer Aided Design of Organic Synthesis (CADOS) [LAURENCO 85]. The more efficient programs, for instance LHASA [COREY 74] or PASCOP [CHOPLIN 78], operate using a knowledge base describing the main organic reactions as chemical transforms (the synthetic methods).

Building up such a knowledge base is a lengthy and expensive task. It is generally built by experts from experimental examples described in the literature. Moreover, the knowledge it contains about the synthetic methods may be heterogeneous because several people have been working on the subject over a long period of time.

During the last few years, several reaction databases have been developed containing an increasing corpus of experimental data in relation to organic synthesis. These databases are a primary source of information which can favourably replace paper documentation in many applications.

These two reasons (difficulty of building up reactional knowledge bases and new computer information about primary sources dealing with reactions) have led us to set off on the GRAMS project (*Génération de Réseaux pour l'Apprentissage de Méthodes de Synthèse*: Network Generation for the Learning of Synthesis Methods) [JAUFFRET 90 1]. In a first approach, we started with an idea of WILCOX and LEVINSON about the automatic classification of reactions [WILCOX 86]. Then, we moved away from this work: in our approach, a greater number of parameters are taken into account and new methodologies are involved.

The GRAMS project has two main parts:

- the design of a suitable system which can 'learn' from several examples of reactions dealing with a similar transformation, the structural and experimental conditions making easier or inhibiting this transformation.

- the design of systems allowing the suitable use of this knowledge in order to i) expect the possible evolution of a set of reactants under given chemical conditions; ii) put forward some synthetic methods generating a given molecule.

The knowledge and operating systems have been developed in our laboratory and they are running as prototypes at the present time.

This paper discusses only the first part of GRAMS (the learning module). We will give more details about the model taking into account the 'reaction' entity in the particular context of a learning process. Some mechanisms of generalisation are then introduced allowing the induction of more general knowledge starting with particular samples (these mechanisms have been presented in a former paper [TONNELIER 90]).Finally, we specify the adaptations we had to perform to the current theory of the learning by induction.

A model of reaction adapted to the context of learning: CGR and CGR+

A chemical reaction is a conversion of a set of reactants into a set of products. During the reaction, the electronic environments of atoms and bonds, and the geometry of molecular structures, are modified. The reaction is best described by the concerned reactants and the reaction conditions (temperature, pressure, solvents, catalysts, pH, etc.). The main features are its mechanism (electronic displacements during the course of the reaction), its kinetic and thermodynamic properties (reaction rate, equilibrium and free energy constants, etc.) [ALLINGER 83].

The usual definition of a chemical reaction may be reduced in our context: taking only into account the 'transformation' aspect, we are mainly interested in the conversion 'reactants-products' and the yield of this transformation, under given experimental conditions. In this context, two kinds of data may be distinguished: i) the reaction core (i.e. the set of bonds and atoms involved in the reaction) defining the reaction class; ii) the different parameters which can modify the yield of the reaction (the structural environment of the core and the experimental conditions).

The learning process by induction is based on the search of similarities and/or differences between the examples. In our project, the comparison between two reactions is a main task and this operation has consequently strong requirements in the mode of the reaction representation. The modification of the reactants is to be described very exactly and without ambiguity. Thus, the representation has to take into account the dynamic character of the reaction and a mapping is to be found between the atoms of the reactants and those of the products: this condition is not fulfilled in the representation commonly used by the chemists: i.e. the developed structural formulae (Fig 1.a). The structural part of a reaction may be represented by a unique graph; thus two connected graphs of reaction allow the immediate comparison of the associated reactions. During the learning process, the present release of GRAMS does not take into account the experimental conditions. Their introduction in the model is not a theoretical problem. We now keep these data under a literal form. The new form is called 'Condensed Graph of Reaction' (CGR) [JAUFFRET 90 2]. It is an extension of the 'Superimposed Reaction Skeleton

Graph' concept described by VLADUTZ [VLADUTZ 88] and re-used by FUJITA
[FUJITA 87].

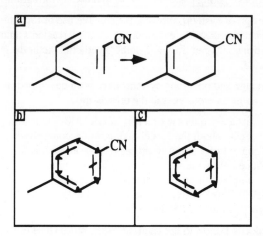

Figure 1: a: A Diels-Alder reaction (usual notation); b: The corresponding Condensed
Graph of Reaction; c: The core of the reaction including all dynamic bonds

The CGR is a connected graph: it superimposes the structures of the reactants and
the structures of the products. Each node corresponds to an atom and each vertex
to a bond. A part of the CGR remains unchanged (unchanged bonds during the
reaction) and another one is dynamic: the reaction core (Fig 1.c). The CGR keeps
the conventional notation of the structure formulae in the invariant part. New
specific notations are necessary to describe the dynamic bonds: a double arrow
represents a new bond (or a bond which increases its bond order by unity); a
crossed line represents a broken bond (or a bond which decreases by unity its
bond order). Most of the chemical reactions may be clearly represented with this
form which is very simple; it is also very easily understood by the chemist (Fig 1.b).

The graph concept fits very well with a computer representation of topological
information. In the CGR, each node has a label which depends on the type of the
corresponding atom and each vertex is labelled according to the nature of the related
bond. However, there are other types of data as the charge of the atoms, their
stereochemistry, etc. In the CGR these data are considered as particular atoms called
'pseudo-atoms': pseudo-atoms are linked to the real atom which is concerned by
such data. The model proposed by VLADUTZ describes only the electronic
rearrangement: our model allows a good computer representation of different kinds
of information in a homogeneous way. The pseudo-atoms may describe any type
of data ; their use is theoretically unlimited. If they are used with dynamic bonds,
they give a good account of the evolution of non-topological parameters during the
reaction. For instance, Figure 2 shows how information concerning the stereochemistry may be included in the CGR in an homogeneous way: in this reaction,
the stereocentre of a halide is inverted (S → R).

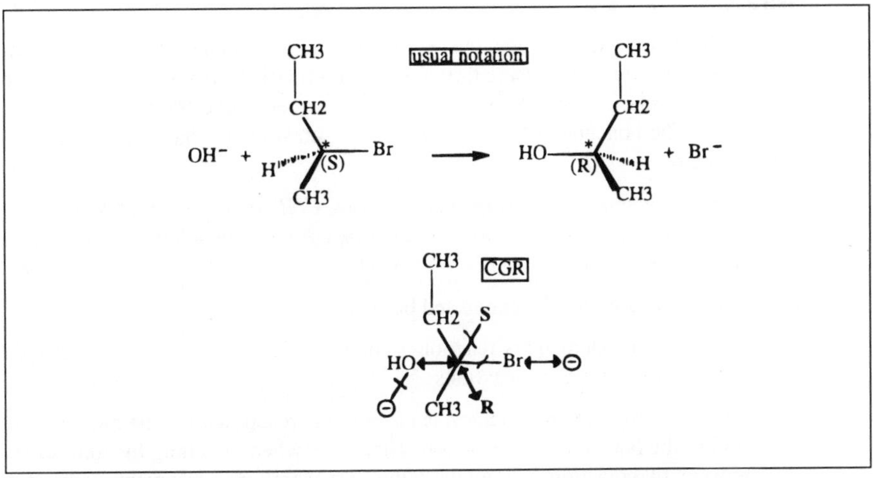

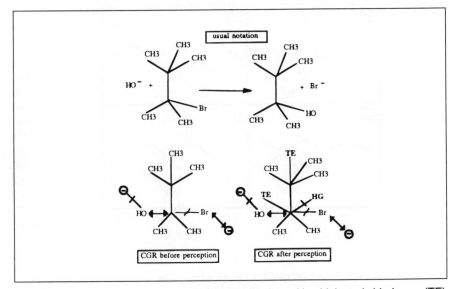

Figure 2: A substitution reaction: usual notation (top) and corresponding CGR (bottom) using the pseudo-atoms formalism (bold)

The model we have just defined is restricted to the description of atom or bond properties. However, the reactional behaviour of a molecule may also be modified if structural 'functional' patterns are present, bringing out some particular features (inductive effects, steric hindrance, nucleophilic or electrophilic characters, etc.). To take these parameters into account during the learning process, the functional patterns are recognised and the corresponding properties are included in the reaction representation, as new pseudo-atoms. The resulting representation is called CGR Plus (CGR+); this is the form used during the learning process (Fig 3).

Figure 3: The perception module has recognised 2 sites with a high steric hindrance (TE) and a substituted halogen (HG + dynamic bond)

The generalisation process

During the generalisation stage, the reactional schemes common to specific examples must be elucidated by using the CGR+ representations. An intuitive method to reveal the similarities between two graphs is to determine their common substructures. The concept of 'maximal common substructure' has been defined to fulfil this aim:

A maximal common substructure between two CGR is a connected sub-graph, common to the two CGR including the reaction core, and which is not included in any other common sub-graph (Fig 4).

Such a substructure must be connected because:

• the core of any elementary reaction is connected, and any reaction may be split into a set of elementary reactions

• the more a substructural pattern is close to the reaction core, the more it may modify the feasibility of a reaction. Therefore, when searching for similarities between two reactions having the same core, isomorphic fragments connected to the two cores by different pathways do not have to be taken into account.

As the perception of maximal common substructures is the essential tool of the induction mechanism, we have built up an efficient algorithm to find them. This algorithm consists of searching the maximal cliques in the compatibility graph of the two CGR+ to be compared [NICHOLSON 87, BARROW 76]. It has been optimised by taking into account different requirements related to molecular graphs and the connectivity constraint. More details have been presented in a former paper [TONNELIER 90].

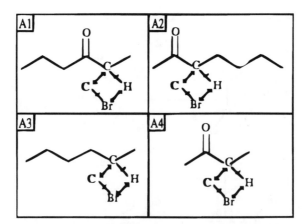

Figure 4: Two alkylation reactions A1,A2 and their two maximal common substructures: A3 and A4

The hierarchical networks

Applied in a recursive way to reactions having the same core, the generalisation process allows the building up of hierarchical networks of reactional schemes (Figure 5).

A hierarchical network is defined by a set of nodes $N = \{n_1, n_2, n_3, etc.\}$ and a set of vertices $A = \{a_1, a_2, a_3, etc.\}$.

Each node corresponds to a reactional scheme (specific reaction for a leaf).

Each vertex from n_1 to n_2 corresponds to the relation 'n_2 is more general than n_1'.

The algorithms already developed to build and to update a network are to be published shortly. These networks are complete (a maximal substructure common to two nodes of the network belongs also to the network), and they are minimum (there is no vertex between two nodes linked by a pathway longer than one step).

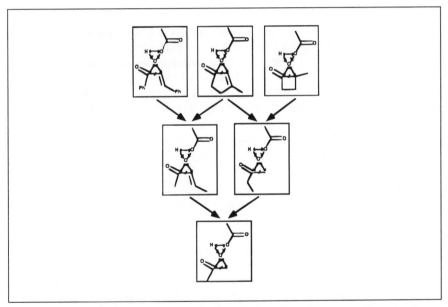

Figure 5: Building up of a hierarchical network, starting from 3 Baeyer-Villiger reactions

Learning by induction within the GRAMS project.

The aim of the project may be recalled at this point. One would like to answer the following questions: 1) Which schemes may be applied to generate a given molecule under specific conditions? 2) Which reactions may occur for a given set of molecules? 3) Which yield can be reasonably expected from a given reaction? Within a class of reactions (i.e. a set of reactions containing the same core), the purpose of the GRAMS system is therefore to induce from examples favourable or unfavourable contexts of the corresponding chemical transform.

The duality 'examples/counter-examples' has no meaning within the GRAMS project. Although there are reactions yielding 100% and other 0%, the yield of a reaction is, in most cases, intermediate to these values due to a non-complete reaction or a competition between several reactions. Consequently, all reactions extracted from literature ("the reaction R has a yield of x% (0<x<100)") will be considered as 'examples' or 'positive instances'.

Some concepts of learning by induction (see for example [DIETTERICH 83]) have to be once more defined, and new ones have to be introduced (an example at the end of the paragraph will illustrate the different terms presented):

The space of models.

For a given reaction class, the space of models is defined as the hierarchical network built up from examples (a network node is a model for all nodes from which it is the generalisation). It can be pointed out that a same formalism (CGR+) allows, at the same time, the description of examples (specific reactions), and the description of concepts (generic schemes of reactions). All network nodes can thus be considered as models.

The discriminating model

In an area where examples and counter-examples may be distinguished, a model is discriminating if it does not recognise (i.e. if it does not generalise) any counter-examples. In the GRAMS project, a 'local' discriminancy may be defined in relation to a yield range; that is, a reactional scheme which does not match any example of reactions will be discriminating if it does not recognise any example whose yield is outside the yield range. The proceeding method is the following:

- Two numerical values may be associated to each network node:
  - the mean yield of the corresponding specific reactions
  - the standard deviation of these yields.

An index of the pertinence of the reactional scheme carried by a node is given by the standard deviation. The less the standard deviation, the more the pertinence. The intuitive idea is the following: if the dispersion is reduced, a yield close to the experimental one will be obtained whatever the reactions generated from this scheme. A model (here a reaction scheme) will be discriminating if its pertinence clue is high (higher than a given limit). The model built up by the disjunction of a set of discriminating models (scheme A or scheme B, etc.) is a discriminating model too, relative to the union of the ranges yields of the examples recognised by the disjunction.

The complete model

The definition used is the common one. A model is complete if it matches all positive instances. There are, in the case of a network, two trivial complete models: one is the root of the network, and the other the set of examples (the model is then built up by disjunction of models).

The consistent model

The most important notion in learning process is the consistent model (i.e. both complete and discriminating). In our case, it would rather be exceptional that an unique network node could be consistent. It would mean that a reactional scheme may be applied whatever the context, and that the yields of the reactions built up on this scheme are close. Actually, the consistency will be reached only by the disjunction of discriminating models, this disjunction having to match the set of examples (by the definition of complete models).

The set of examples is a consistent model of minor interest (nothing is learned). The concept of optimal model is therefore introduced.

The optimal model

The optimal model is the set of discriminating nodes which cannot be generalised into another discriminating node. Such a model is the best which can be expected from a system such as GRAMS: It is a discriminating model (because obtained from the disjunction of discriminating models). It can easily be proved that it is also complete. Then, it is consistent.

The optimal model describes the relations (structural/experimental conditions → yield of the reaction) with a minimal number of rules. Thus, it permits the concise description of a chemical transform. The relative importance awarded to the precision or to the concision may be controlled by tuning the pertinence limit. The lower the limit, the more concise the model (less nodes will contribute to the disjunction). But the learned knowledge will become less precise, and so will be the results given by the systems which will handle that knowledge. Only the current use of the program will allow the adequate definition of this parameter.

A simple example of network which yields between 10% and 90% (Fig 6) will illustrate these notions. The dispersion of yields is important for the root (n9) because dominated by all the leaves. Its interpretation is difficult and the associated model nearly not pertinent. The same is happening to node n5. But, for nodes n3, n7, n11, and n12, the dispersion is low. According to GRAMS criteria, the associated models are discriminating. The set of models [n3, n12] is therefore a discriminating model too. Since the model recognises all the examples, the set is complete (and therefore consistent). Lastly, nodes n3 and n12 can not be generalised as discriminating models (because neither n5 nor n9 are discriminating). [n3, n12] is thus the optimal model. Taking the examples into account, it can be deduced that the chemical transform may be described by schemes n3 and n12 (n3 representing an unfavourable context whereas n12 is a favourable one). Since this virtual network has only 6 leaves, this analysis is surely rough.

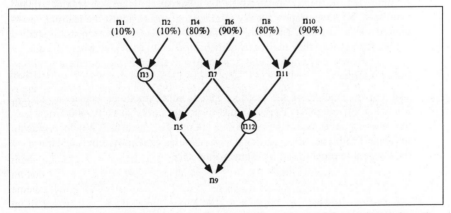

Figure 6: nodes n3 and n12 give a correct description of the transform, starting from the given examples

Reliability of a reactional scheme

As earlier defined, the pertinence is not sufficient to assure the chemical value of the model. For instance, two very different contexts of a same chemical transform will only have very general common schemes. According to the GRAMS approach, a common scheme will be considered as pertinent if the yields of the examples of reaction corresponding to these contexts are close. It means that for all reactions built up on this scheme, the system will associate a potential yield neighbouring the yields of the examples. But in most cases this statement is chemically wrong. Such an over-generalisation results from a too limited number of examples. In the above example, a third reaction recognised by the same scheme, but yielding in a totally different range, would have led to a non-pertinent scheme.

A reliability index is associated with each network node to overcome this inadequacy. At the moment, only the number of examples recognised by a scheme is taken into account for the estimation of this index. The reliability of some examples is not questioned. A reactional scheme will be considered as more reliable if it generalises more specific reactions, because more examples assert the validity of the scheme. This process will be later improved by considering the chemical similarity of the examples.

The reliability notion is only useful in addition to the pertinence notion, though the reliability index may be defined for each network node. The reliability index may be used in two different ways by the systems using the learned knowledge: either to balance the accuracy of the proposed results, or to select among the reactional schemes of the optimal model, the ones resulting from a sufficient number of examples.

Conclusion

The first part of the GRAMS system dealing with the learning of synthetic methods has been described in this paper. The driving force of this project is the bottleneck represented by the building up of knowledge bases needed by most of CADOS programs. An adequate formalism has been elaborated to describe reactional knowledge. Furthermore, sophisticated algorithms have been realised to generalise examples of tested chemical reactions. A theoretical work about the learning process by induction has been carried out in the concrete and complex area of organic synthesis. We have proposed some original concepts to adjust the learning methodology to this domain. The interest of the introduced concepts exceeds the studied application. In fact, they may potentially be applied to each area where the duality examples/counter-examples can be replaced by a reliability measurement (continuous or discrete).

The implementation of the learning and handling modules began on a DEC MicroVAX II computer. These modules have been ported to a HP 9000/730, where they are now developed. They are written in C language with an XWindows graphic interface. Prototypes are operational for both modules. The system is under tests at the moment in order to sharpen some parameters.

Bibliography

[ALLINGER 83]: Allinger N.L., Cava M.P., De Jongh C., Johnson C.R., Lebel N.A. and Stevens C.L., 'Chimie Organique', McGraw-Hill, Paris, 1983

[BARROW 76]: Barrow H.G. and Burstall R.M., 'Subgraph Isomorphism Relational Structures and Maximal Cliques', *Information Processing Letters*, 1976, **4**, 83-84

[CHOPLIN 78]: Choplin F., Marc R., Kaufmann G. and Wipke W.T., *Journal of Chemical Information in Computer Science*, 1978, **18** (2), 110

[COREY 74]: Corey E.J., Howe W.J. and Pensak D.A., 'Computer Assisted Synthesis Analysis Methods for Machine generation of Synthetic Intermediaries involving Multistep Look-Ahead', *Journal of American Chemical Society*, 1974, **96**, 7724

[DIETTERICH 83]: Dietterich T.G. and Michalski R.S., 'A comparative review of selected methods for learning from examples' in 'Machine Learning', R.S. Michalski, J.G. Carbonnel and T.M. Mitchell (Eds), Tioga, Palo Alto, CA, 1983, 41-81

[FUJITA 87]: Fujita S., 'Structure-Reaction Type' Paradigms in the conventional methods of describing Organic Reactions and the concept of Imaginary Transition Structures Overcoming this Paradigm', *J. Chem. Inf. Comp. Sci.*, 1987, **27**, 120-126

[JAUFFRET 90 1]: Jauffret P., Hanser T., Tonnelier C. and Kaufmann G., 'Machine Learning of Generic Reactions: 1. Scope of the Project; the GRAMS Program', *Tetrahedron Comput. Methodol.*, 1990, **3** (6), 323-333

[JAUFFRET 90 2]: Jauffret P., Tonnelier C., Hanser T., Kaufmann G. and Wolff R., 'Machine Learning of Generic Reactions: 2. Toward an Advanced Computer Representation of Chemical Reactions', *Tetrahedron Comput. Methodol.*, 1990, **3** (6), 334-350

[LAURENCO 85]: Laurenco C., 'Synthèse assistée par ordinateur', Thèse de Doctorat d'Etat, Strasbourg, 1985

[NICHOLSON 87]: Nicholson V., Tsai C., Johnson M. and Naim M., 'A Subgraph Isomorphism Theorem For Molecular Graphs', *Studies in Physical and Theoretical Chemistry*, 1987, **51** 226-230

[TONNELIER 90]: Tonnelier C., Jauffret P., Hanser T. and Kaufmann G., 'Machine Learning of Generic Reactions: 3. An Efficient Algorithm for Maximal Common Substructure Determination', *Tetrahedron Comput. Methodol.*, 1990, **3** (6), 351-358

[VLADUTZ 88]: Vladutz G., 'Joint Compound/reaction storage and retrieval and possibilities of a hyperstructure-based solution' in 'Chemical Structures', W.A. Warr (Ed.), Berlin Heidelberg, 1988, 371-384

[WILCOX 86]: Wilcox C.S. and Levinson R.A., 'A Self-Organised Base for Recall, Design and Discovery in Organic Chemistry' in 'Artificial Intelligence Application in Chemistry', T.H. Pierce et B.A. Hohne (Eds), 1986, 209, 230

Neural Networks: a new computational paradigm with applications in chemistry

David W. Elrod and Gerald M. Maggiora

Computational Chemistry, Upjohn Laboratories, Kalamazoo, MI. USA

Abstract

Computational neural networks are a new computational paradigm, inspired by the brain's massively parallel network of highly interconnected neurons. The interest in computational neural networks (CNN) derives mainly because they offer a new general paradigm for model building and function approximation, and not because of their relationship to models of brain function. Neural networks are considerably different from other computational methods in that functional relationships are not programmed in but are learned directly from data. This ability to adapt to new data, along with the 'model-free' nature of neural networks allows them to excel at constructing very general, complex, nonlinear mappings between sets of input-output pairs, even with noisy data. In fact, CNNs can construct mappings of arbitrary accuracy for all reasonably well-behaved functions. In addition to their mapping ability, neural networks are also useful for pattern recognition/ classification and self-organised clustering.

A survey of the rapidly increasing number of applications of neural networks in chemistry will highlight some of their properties and limitations, and show what their impact may be in computer-aided chemistry.

Keywords: Computational neural networks, artificial neural networks, review, chemical applications.

1. Introduction

The way in which the brain processes and stores information has fascinated scientists for many years, and this fascination has led to numerous attempts to simulate the brain's functions by artificial intelligence (AI) procedures implemented on digital computers. All of these attempts have to some extent failed. While the reasons advanced for the failures are many, it is perhaps the fact that the brain's ability to carry out massively parallel computations on its highly interconnected network of neurons is not incorporated naturally into the AI procedures. And it is the massively parallel architecture that largely confers upon the brain, among other things, its ability to learn and to carry out sophisticated pattern recognition. Today, only a small number of digital computers possess highly parallel architectures, and even those that do not remotely approach the level of connectivity found in the brain.

Biologically, the brain's connectivity is achieved through its neurons, which possess a number of structural features conducive to supporting a high level of interneuron connectivity. Neurons possess a cell body or soma from which emanate a set of root-like extensions called dendrites and a single fibre called the axon, which terminates in a series of smaller axonal fibres. Signals are transmitted between neurons from the axon of one to a dendrite of another across a synaptic junction, the strength of the transmitted signal being determined by the biochemical state of the particular synapse. As each neuron possesses a number of dendrites as well as an axon which ends in multiple branches, it is clear that a high level of interconnectivity can be achieved in the brain. Another important feature of neurons is that they exhibit threshold behaviour; that is, a neuronal signal is not generated until the 'sum' of the incoming signals to the neuron exceeds a value which is determined by its biochemical state.

Even though the structure of neurons and their interconnectivity was known before the end of the nineteenth century, it was not until the work of McCulloch and Pitts in 1943 that a serious attempt was made to model features of the brain mathematically.[1] While these investigators showed that arbitrary logical functions could be implemented on a set of interconnected 'mathematical' neurons that individually exhibited all-or-nothing threshold behaviour, they did not provide a means for the network of neurons to learn, that is, a mechanism whereby interneuron connections could be established or terminated. In 1949, Hebb developed a learning scheme, now called Hebbian learning, that provided a practical procedure for neural networks.[2] From that point until the early sixties most work involved the development of a variety of interesting machines designed to emulate various aspects of the information processing functions of the brain. From then on until today much of the work has involved simulations of neural networks on sequential computing machines, although a number of electronic and electro-optical devices as well as several massively parallel computers have also been constructed for implementing functional neural networks.[3]

In this latter period, interest solely in emulating the brain began to be displaced by an interest in learning machines in general. This freed scientists from the constraints prescribed by a need to model the brain explicitly and led to a plethora of neural-network paradigms, which nevertheless still retained many 'brain-like' features. To distinguish neural networks that are designed to model the brain explicitly from those that retain only some of the brain's perceived computational characteristics, the latter are usually referred to as *computational neural networks* or *nets* (CNN). Computational neural nets are also called connectionist systems, parallel-distributed-processing systems, artificial neural networks, adaptive systems, adaptive networks, neurocomputers, artificial neural systems (ANS), learning machines, and generalised perceptrons, to name a few.

Kosko [4] has pointed out that CNNs can be considered as *model-free* mapping devices in so far as the functional form of the mapping need not be specified explicitly, in contrast to the situation in both linear and non-linear regression methods. While this may be true technically, it omits the fact that the network topology, *i.e.*, the number of hidden layers and the number of nodes in each layer, as well as the type of transfer function must at some point be specified, and this may be construed, in some sense, as being analogous to choosing a model. This model-free character would appear to provide an advantage to CNNs in cases, commonly encountered in chemistry, where complicated input-output relationships are an inherent feature of the system or systems under investigation.[5] To investigate such systems, however, is difficult due to the problem of finding suitably complete data-sets.

In Section 2 a number of general characteristics of CNNs are considered. Special emphasis is given to multilayer, feed-forward neural networks, which are currently the most popular, appearing in more than 80 percent of all chemical applications. These neural networks are sometimes incorrectly called back propagation networks, due to the fact that training in such networks is usually carried out by the back propagation-of-errors learning algorithm. That terminology should, however, be avoided, since it describes the most common training method but does not specify the network architecture. Section 3 provides an overview of chemical applications of CNNs and includes brief discussions of important examples in each of the ten application categories. The final section, Section 4, considers a number of important issues that bear upon both fundamental aspects of CNNs as well as their application to problems in chemistry.

2. Characteristics of computational neural networks

Although based loosely on analogies to the brain, interest in neural networks thus derives not from their ability to model the brain, but rather from their ability to treat a diverse set of problems using a highly-distributed parallel computer architecture. [3,6-9] Neural networks are made up of collections of highly-interconnected, but relatively simple processing elements (PEs); each interconnection has an associated weight. The strength of the connection between two PEs, which is due to the biochemical state of the synapse in biological neurons, is determined by the connection weight, or simply the weight, in CNNs. In general, CNNs are distin-guished by their network paradigms, which define the nature of their PEs, the network topology or pattern of connectivity among the PEs, and the learning method. There are two general categories of learning methods, unsupervised and supervised.[3,6-9] In unsupervised learning the network itself determines the ap-propriate set of weights, while in supervised learning weights are determined such that the error between a set of training data and network predictions is minimised. Thus, learning in CNNs is any procedure for altering the weights in response to the input of new information. Unlike traditional computers that store programs and data separately, CNNs store information in the distributed pattern of their interconnec-tions and values of the associated weights.

Of the many CNN paradigms that have been investigated, we will focus our attention here on multi-layer, feed-forward nets, also called generalised perceptrons, with either back propagation or stochastic supervised learning.[10,11] Hopfield nets, which are sometimes considered as being largely responsible for bringing about the present resurgence in neural net research and applications, will not be discussed explicitly.[12,13] While a number of chemical applications have been carried out with this type of CNN, most of the activity has centred around the prediction of the tertiary structure of proteins.[14] Due to the nature of Hopfield nets it is not expected that they will play as ubiquitous a role as other paradigms in future chemical applications. Kohonen Self-Organising Maps [15] (SOM) will only be briefly mentioned in the survey of chemical applications of CNNs. A compre-hensive description can be found in the work of Professors Zupan [16] and Gasteiger.[17]

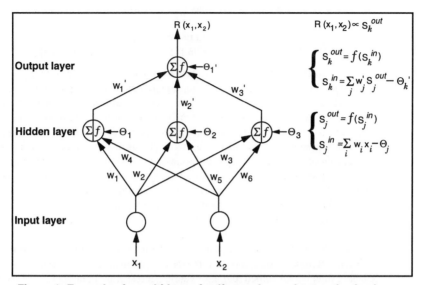

Figure 1: Example of a multi-layer, feedforward neural network, also known as a generalised perceptron

Generalised perceptron, feed-forward CNNs have been successfully applied to a wide variety of problems and are the most extensively studied to date. A typical example of such a CNN is depicted in Figure 1. As shown in Figure 1, each PE performs two operations, a summation denoted by 'Σ', followed by a function evaluation denoted by 'f'; f is called a transfer function (or activation function) and is often either a sigmoid function or the closely related hyperbolic tangent, *tanh*, function shown in Figure 2. Each PE also possesses a threshold or 'bias weight,' generally denoted Θ_t, which when combined with a constant, unit signal, effectively shifts the transfer function along its abscissa (cf. Figure 1).

CNNs can be viewed in two ways, either as classifying or as mapping devices. Generally, but not always, classification represents a less stringent, although non-trivial, test of a CNN's performance than function mapping. This follows from the fact that in the latter case the value of the function over the domain of interest must be predicted, while in the former case only whether the function value lies above or below some threshold is required. Nevertheless, it has been shown by numerous

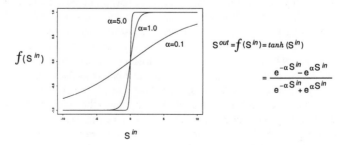

Figure 2: Example of a non-linear transfer function *tanh*

workers that generalised perceptrons can accurately represent the mappings of essentially all reasonably well-behaved functions. [18-22] The proofs, however, are only proofs that a solution exists and unfortunately, do not indicate precisely how one should produce a CNN that can, in fact, carry out the desired mapping.

Figure 1 depicts the network architecture of a typical three-layer feed-forward net (*N.B.*, the input layer is sometimes not counted as it merely passes on the input values to the next layer in the network). Feed forward networks without hidden layers are generally called *perceptrons* and thus, feed-forward networks with hidden layers are sometimes called *generalised perceptrons*, and that is the terminology that will be used in the remainder of this work. While many problems can be handled by perceptrons, some extremely simple ones cannot. The classic example, as pointed out by Minsky and Papert [23], is the inability of perceptrons to solve the simple XOR problem. It is usually assumed that this inability of perceptrons to solve the XOR problem was primarily responsible for the decline in popularity in the 1970s of neural network methods. Although generalised perceptrons can easily solve this problem, a satisfactory procedure for training them was not available until the landmark paper by Rumelhart, Hinton and Williams [24], which was published in 1986. In that paper, the authors describe the now famous back propagation-of-errors learning rule, which helped revitalise neural net research. Interestingly, Werbos, in his 1974 Ph.D. thesis [25], essentially described the back propagation learning rule as a technique for function optimisation, but due to the limited distribution of his work, it was not recognised until much later.

As shown in Figure 1, each layer of the network in a generalised perceptron consists of a set of nodes, indicated by circles, joined by weighted (w_i), uni-directional connections indicated by arrows. There are no connections between nodes within a given layer in this type of network architecture. The values of the input variables, x_1 and x_2, are 'passed through' the input nodes without change. The nodes in all other layers are called processing elements (PE) and carry out both a summation of the incoming signals, denoted by 'Σ,' and an evaluation of the resultant sum by a non-linear transfer function, denoted by 'f.' The detailed form of these functions is shown to the right of the figure. In addition to the weighted inputs summed by each PE, an additional threshold or bias term, denoted by Θ_t, is also added to the sum. The final output, $R(x_1,x_2)$, is then obtained by an appropriate scaling of the value of the transfer function, S_k^{out}, in the output layer. Scaling is required as the output range of most transfer functions is given by [0,1] or [-1,1], as seen for the hyperbolic tangent transfer function, *tanh*, depicted in Figure 2.

The ability of CNNs to learn, *i.e.* adapt to new input data, distinguishes them from most other computational paradigms. Learning in CNNs consists of appropriately modifying the network weights in response to a given set of inputs and can be classified as either *supervised* or *unsupervised*. In the former case, learning, or *training*, is based upon a direct comparison of the network's output with its input, where the outputs represent 'correct' network responses. In the latter case, the learning goal is not generally specified in terms of correct network responses, rather the network is expected to use correlations among the input data to create categories which reflect these interrelationships. In either case, memory is embodied in the set of weights which defines the network. Thus, there is a clear distinction between memory in CNNs compared to memory in typical digital computers, even those which support highly-parallel computation.

Generalised perceptrons are trained by supervised learning methods. Currently the most popular training procedure is back propagation. In this approach, an error function is defined that determines, for the given collection of weights, the difference between the desired network output and the output produced by the net for a set of inputs, *i.e.* the training set.[26] During the learning process the weights are incrementally adjusted until the error function reaches a minimum, which is usually a local minimum. Finding the global minimum by such a procedure is very difficult and is generally not possible except in very special circumstances. The difficulty is a result of the fact that back propagation training, like other gradient descent methods, makes changes in the weights proportional to the gradient, or first derivative of the error function. The more the error function decreases, the larger the gradient becomes and consequently, the greater the weight changes. These weight changes can only be made in the direction that leads towards lower error. Once the error function is in a valley on the error surface, the back propagation algorithm will find the bottom of the valley, but it cannot climb over a ridge to a lower valley. The weight adjustment is carried out sequentially beginning with those weights closest to the output nodes. The error correction or adjustment of the weights is then 'propagated,' layer by layer, back towards the input layer. Thus, the input signals are propagated forward and the 'error signals' are propagated back towards the inputs, hence the use of the terminology back propagation (BP) learning to describe the training process. Other generalised perceptron learning procedures that avoid some of the problems that BP has in getting trapped in local minima have been implemented, but BP remains the most widely used method to date.[27]

Numerous neural network architectures support unsupervised learning, and the number of applications in chemistry based upon these networks is increasing steadily. However, since supervised learning methods, like BP, have been the most widely used type of neural networks for chemical applications, they will be the focus of this discussion. There are, however, a number of excellent detailed treatments of these unsupervised neural net methods available.[6-9,28]

Generalised perceptrons can be viewed in a number of ways. Two of the most useful are as devices that can implement either discriminant functions or generalised mappings. The former is important in classification problems, while the latter is important in property prediction and in modelling both static and dynamic systems, to name just a few. Lippmann has presented a very clear discussion of generalised perceptrons from the perspective of discriminant functions. [9] He shows that the presence of hidden layers in generalised perceptrons allows them to construct boundaries of great complexity, a feature that is missing in perceptrons, which can only construct linear discriminant functions. In fact, it is this latter limitation which prevents perceptrons from solving the XOR problem.

Perhaps the most powerful feature of generalised perceptrons is their ability to construct very general mappings between sets of input-output pairs. In fact, generalised perceptrons can construct very complex, non-linear mappings of arbitrary accuracy for all reasonably well-behaved functions.[8,10,29] These mappings can be classified as either *auto associative*, if the input-output pairs are identical, or *heteroassociative*, if they differ. Networks with these types of behaviour are called auto associative and hetero associative, respectively.

Advantages	Disadvantages
Massively parallel	Slow to train
Noise tolerant	Difficult to explain predictions
Learn by example	Non-algorithmic
Improve with training	Local minima problem
Non-algorithmic	Overfitting/overtraining
Associative memory	Data dependent
Content-addressable memory	Representation dependent
Distributed representation	Requires vector data representation
Pattern recognition	May extract incorrect features
Pattern classification	May generalise in valid but undesired ways
Non-linear mapping	
Feature extraction	
Construct rules from data	
Generalisation	
Model Free	
No assumptions about data distribution	
Global	
Arbitrary complexity	
Adaptive	
Non-parametric	

Table 1. Neural Network Characteristics

In addition to the features noted above, CNNs possess a number of other features that make them potentially suitable for a variety of tasks that are difficult to address with more traditional computer architectures. These characteristics are shown in Table 1. It is also appropriate to comment at this point on one of the distinguishing features of CNNs, namely their massively parallel architecture. The operation of a neural network is inherently parallel, with each PE processing its own inputs and outputs. However, in most work to date, the CNN was simulated on a conventional computer with a single, sequential processor. When the appropriate network design has been worked out for a problem, then scaling up that solution by implementing the CNN on a multi-processor computer will be much easier than rewriting traditional computer programs to make them run on parallel computers.

3. Applications of computational neural networks in chemistry

Neural network applications have been growing at a rapid rate in general, while applications in chemistry had been lagging somewhat behind until chemists learned how to apply this 'new' technology. It now appears that chemists are beginning to exploit this methodology to solve a wide range of chemical problems. A search of The American Chemical Society's Chemical Abstracts (CA) database [30] on STN

Chemical Applications of Neural Nets

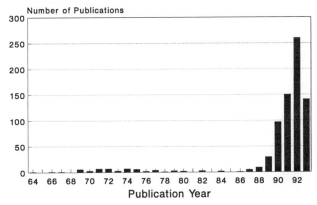

Total • 754 Publications to Aug 1993

**Figure 3: Histogram showing the number of chemical applications of CNN papers
published per year from 1964 to the present**

International, from 1967 through August 1993 found 598 papers on CNNs. An additional 128 chemical applications of CNNs were found in the INSPEC engineering database [31] on STN, eleven applications were found in the COMPUSCIENCE database [32] on STN, thirteen dissertations were found in the DISSABS database[33] on STN, and another four papers were found by examining the current literature directly. This survey of 754 papers is undoubtedly incomplete as new works are published weekly. Figure 3 shows a plot of the number of chemical applications of CNNs as a function of year. From the figure, it is clear that more than half of the publications have appeared within approximately the last year and a half (402 papers or 53%), and nearly three-fourths occurred within the last two and a half years (552 papers or 73%). The earliest papers systematically applying neural networks in chemistry appear to be those of Jurs, Kowalski, Isenhour, and Reilly [34] at the University of Washington in 1969, when the first of a long series of papers appeared in the journal *Analytical Chemistry* under the series heading 'Computerised Learning Machines Applied to Chemical Problems.' Four earlier papers appeared from 1964 to 1968, but they were of a more theoretical than practical nature. Ironically, 1969 was also the year that Minsky and Papert's book *Perceptrons* [23] appeared. The small blip on the graph for 1969–73 represents work with linear learning machines [35] (perceptrons), whose limitations were soon noted. A large rise in the number of papers followed publication of the BP learning algorithm [24] in 1986, which revived the field and led to the current resurgence in the development and application of CNNs in many fields, including chemistry.

Neural network applications in chemistry can be divided into roughly ten categories, as shown in Table 2. These ten categories are somewhat arbitrary, and in some cases may have considerable overlap, but they serve to categorise the various areas of chemistry where neural networks have seen application. There is a very large body of literature on engineering and process control applications for neural nets. Papers that did not specifically deal with control of chemical or biochemical processes, or that were not abstracted in the CAS CA file were not included in the process control category. A number of references describing the chemistry of materials for constructing neural network chips and devices were also not included. An overview of such a large number of papers must, by necessity, be selective. Also, due to space limitations, it is not possible to discuss all the important work in each area. In many instances, there were several equally important articles, but only one, usually the most recent, was included in the summaries that follow. In the following sections, each category will be described briefly, emphasising the range of types of applications, trends, important new methods, and the neural-network paradigms used, as filtered through the perspective of an organic/medicinal chemist.

Category of applications	Number of papers per category
Reviews, theoretical studies, & miscellaneous	107
Chemical process control & chemical engineering	173
Analytical chemistry	139
Protein & polymer structure & analysis	52
Biomolecular informatics	35
QSAR & pattern recognition	54
Spectra-structure correlation	67
Property prediction & parameter estimation	26
Reaction & reactivity prediction	11
Nuclear chemistry & nuclear power	90

Table 2. Categories of published neural network applications in chemistry

Thirteen of the publications are actually dissertations, starting with P Jurs [36] in 1969, who studied linear learning machines for mass spectral interpretation at the University of Washington. After a long hiatus, three chemical engineering dissertations that applied neural nets appeared in 1991: (2) S. Roat's (U. Tennessee) application of neural networks for nonlinear optimal control of chemical processes, [37] (3) D. Haesloop's (U. Washington) system identification and control using neural networks, [38] and (4) N. Bhat's (U. Maryland) use of BP networks for control of dynamic chemical processes.[39] The other nine dissertations were in the areas of analytical (3, total = 4), bioinformatics (1), process control (1, total = 4), reaction prediction (1), and spectroscopy (3), indicating that neural networks are being recognised as areas of current research interest. It is also worthy of note that one of the pioneers of applying neural nets (linear learning machines)in chemistry, Peter Jurs, is again working in this area and he presented three neural net application papers at the August 1992 American Chemical Society meeting in Chicago. [40] Zupan and Gasteiger have also recently done a lot of work on applying neural networks in chemistry [17], and in 1993 published a book, *Neural Networks for Chemists: A Textbook*, which is both an introductory text as well as a survey of applications.[41]

3.1 Reviews, theoretical studies and miscellaneous applications

One hundred and seven papers are found in this category. Included here are reviews, new training methods, general or theoretical studies on neural networks, comparisons with statistical methods, and unusual applications that do not readily fall into the other classes. Two reviews are of particular interest. Zupan and Gasteiger [17] provide an excellent overview of CNN paradigms and describe a number of chemical applications in terms of problem type; and Elrod and Maggiora [42] surveyed some 339 papers on neural net applications in chemistry that had been reported by September 1992. Lacy [43] edited a *Symposium in Print* that listed some of the early CNN applications in chemistry and also gave a brief review as an introduction to the area. Schmuller suggests possible CNN applications in environmental chemistry.[44] Wythoff has written a tutorial on BP nets.[45]

Several papers detail how neural networks can be used to implement both linear [46] and nonlinear [47] partial least squares and also nonlinear principal components analysis. [48] The advantage of neural networks in these methods is their ability to model nonlinear relationships while attaining robust generalisation properties. De Veaux *et al* [49] compared neural nets with multi-variate adaptive regression splines (MARS) and concluded that in most cases MARS was more accurate and faster. The integration of fuzzy logic and neural networks is being explored by a number of researchers. Two recent papers are on fuzzy linear interpolating networks [50] and on implementing a fuzzy expert system in a neural network.[51] Kateman and Smits [52] give criteria for validation and evaluation of neural nets in an attempt to obtain 'coloured information from a black box.'

A novel method for the display of multi-variate physicochemical properties of biologically active molecules was reported by Livingstone, Hesketh, and Clayworth.[53] They used an auto-associative BP network to reduce multidimensional data to two dimensions, thus enhancing the visualisation of relationships within the data. This type of application will likely become more important in the future as chemists discover the potential of neural networks for clustering and dimensionality reduction of high-dimensional chemical data. Zupan and Gasteiger [17] presented two examples of using Kohonen SOM's for reduction of dimensionality of high dimensional data in their 1993 review, and an example by Rose *et al* is in the QSAR section (Section 3.6). Reibnegger *et al* [54] reported on self-organising networks for clustering in clinical chemistry applications. Included among the miscellaneous applications are: neural network solution of the Schrodinger equation [55], using CNNs to capture the knowledge of expert chemical plant operators [56], designing fermentation media [57], catalyst design [58], and for using neural networks for efficient browsing through databases.[59]

3.2 Chemical process control and chemical engineering applications

The largest group of applications of CNNs in chemistry is in the area of process control and chemical engineering, where there are 173 papers. Chemists and chemical engineers who deal with chemical and biochemical processes are used to describing and understanding those nonlinear processes in terms of mathematical models. The majority of the neural networks used were multi-layer, feed-forward nets, trained with some form of BP since many of the applications in this area are concerned with real-time diagnostics and predictive control for nonlinear dynamic chemical processes. The earliest application of CNNs in chemical engineering was in 1988, where Hoskins and Himmelblau [60] gave an outline and a simple example of how chemical engineering knowledge can be represented in a CNN. A recent issue of *Computers in Chemical Engineering* [61] is devoted to reviews of the use of neural networks for process modelling, process fault diagnosis, process variable estimation, and process control. Bugmann, Lister and Von Stockar [62] found that neural networks were useful for characterising bioreactor processes where it was difficult to explicitly model desired properties from the measurements that were able to be made. They found that the CNN could approximate unknown functional relationships from suitable examples without needing to specify the form of the mapping. Expert systems and knowledge-based control systems are finding greater utility in process fault diagnosis and control, [63,64] but there is a bottleneck in the rate at which process knowledge can be acquired. The problem is compounded because of frequently changing process conditions, which require constant updating of the process knowledge. In this area, CNNs seem to be useful because they can extract knowledge patterns directly from plant or process operations data and can be easily updated as new operations data is obtained.

3.3 Analytical chemistry applications

Analytical chemistry is the area with the second largest number of CNN papers. Of the 139 papers in this category, the early papers mainly employed perceptrons or linear learning machines, while more recent publications generally employ generalised perceptrons trained by back-propagation. Roughly one third (46 of 139) of the papers describe the application of CNNs for calibrating and deconvoluting spectrometer data, which was reviewed by Bos *et al.* [65] Another area of concentration where neural networks have seen application in analytical chemistry is in the detection, quantitation, and identification of odours and vapours from semiconductor gas sensors where there are 20 papers, examples of which are found in the work of Chang *et al.* [66] and Nakamoto *et al.* [67], along with many others. Another area where neural networks are beginning to make an impact is in clinical analytical chemistry. A recent example is the non invasive diagnosis of coronary artery disease from heart sounds using CNNs, reported by Akay.[68]

The rest of the analytical applications vary from geology, [69] predicting the water content of cheese, [70] pattern recognition of chromatographic data, [71] particle shape classification, [72] and attempts to develop intelligent analytical instruments. [73] Freeman used neural nets to shape NMR pulses for selective excitation high resolution NMR.[74] Casale and Watterson [75] report on using CNNs as a pattern recognition method in forensic analysis for identifying cocaine signatures, and for linking specific cocaine samples with the bulk batches of drug to help build conspiracy cases. A recent paper by Otto, George, Schierle, and Wegscheider [76] suggests combining fuzzy logic, as a model of an analytical chemist's reasoning process, with neural networks for automatic knowledge acquisition. The goal of this integrated approach is to develop intelligent systems for the automated qualitative analysis of spectroscopic data.

3.4 Protein and polymer structure and analysis applications

Fifty-two papers were found in this category; the majority are attempts to predict the secondary structure of proteins, using feed forward networks. The application of CNNs for protein analysis and structure prediction was recently reviewed by Hirst and Sternberg.[14] Stolorz, Lapedes and Xia [77] found that a CNN did no better than other statistical methods, and concluded that the primary amino acid sequence of the protein does not contain sufficient information to improve on the secondary structure predictions made by statistical methods. Muskal and Kim [78] had more success in predicting secondary structure when they linked two general-ised perceptrons networks in series and used amino-acid composition and other data for inputs. The tertiary structure of proteins has been approached by Wolynes and coworkers [79] employing associative memory Hamiltonians to recognise the folded tertiary structure of proteins. Ferran and Ferrara [80] clustered proteins into functional families using a Kohonen self-organising map.[15] Four of the papers included in this group describe CNN methods for mapping the potential-energy surfaces of synthetic polymers, but may have applicability for proteins as well.[81] A paper by Bohm [82] suggests that CNNs may help overcome some of the lack of knowledge about determinants of protein folding by extracting relevant information from known protein structures. Bohr *et al* [83], describe a method for folding protein backbones from distance inequality information.

3.5 Biomolecular informatics applications

Related to, but distinct from, the preceding category is the growing area of biomolecular informatics. This field is concerned with finding patterns in biological sequences and relating those sequence patterns or motifs to their biological function. The review by Hirst and Sternberg [14] analysed nine of the 35 publications in this area. The majority of the papers focused on using perceptrons for finding promoter sites or binding sites in DNA.[84] Some of the earliest work on finding translation initiation sites in *E. Coli* was done by Stormo and collaborators [85] in 1982. Recently Arrigo, Giuliano, Scalia, Rapallo, and Damiani [86] used a Kohonen self-organising map [15] to identify a new sequence motif in the human insulin receptor gene. Four of the papers deal with protein sequences, with the most recent publication by a group in Hungary.[87] Claverie and Sauvaget have developed a portable software package, [88] WOBB.C, for implementing the perceptron para-digm for defining and recognising ambiguous sequence motifs in either protein or nucleic acid sequences. Frishman and Argos [89], reported a method employing numerous neural nets in a three step process to recognise distantly related protein sequences using conserved motifs.

3.6 Quantitative structure-activity relationships and pattern recognition applications

Quantitative structure-activity relationships (QSAR) have been developed using CNNs to map sets of chemical-structure descriptors to activities in biological systems. The 54 papers found in this area show that the use of CNNs is gaining acceptance as a reasonable alternative to statistically developed models. A major problem seems to be the selection of the proper input variables for the neural net. Wikel and Dow [90] describe a method for variable selection in QSAR by analysing the pattern of weights in the trained neural network. Several authors, including Andrea [91], Aoyama [92], Tetko [93], Maggiora [94], and Livingstone [95] have analysed a number of issues related to the size of data sets and the reliability of the QSARs derived from CNNs. The internal models constructed by the CNNs can be interpreted as mappings or discriminants built up of linear combinations of the transfer functions which can be considered as basis functions. Nearly all of the neural networks used in QSAR applications were of the feed-forward type, trained by BP. Examples include prediction of biodegradation of organic benzene deriva-tives, [96] development of a QSAR for 256 dihydrofolate reductase inhibitors, [91] structure-odour relationships of nitrobenezenic musks, [97] and Hussain's [98] method for formulation optimisation in pharmaceutical product development. Cambon and Devilliers [99] applied CNNs to predicting structure-biodegradability relationships, and Weinstein *et al.* [100] predicted the mechanism of action of cancer drugs from their pattern of activity against 68 malignant cell lines. Rose, Croall, and MacFie [101] used an unsupervised learning method, Kohonen SOM topology-preserving mapping, to reduce a multidimensional matrix of physicochemical property data for some antifilarial compounds to a two dimensional representation. In this case principal components analysis failed to give interpretable clusters, presumably because of nonlinear features in the dataset.

3.7 Spectra-structure correlation applications

Two different types of applications have employed neural networks for the correlation of spectra and chemical structure. In 59 of the 67 papers published to date in this area the input to the network has been spectral features and the desired output was the class of compound or structural fragments present in the molecule that produced that spectrum. The eight remaining papers involved ^{13}C NMR spectral prediction based upon the structural fragments or molecular environments present in the molecules. Three papers employed Kohonen SOM's while the rest of the papers used BP or linear learning machines, which was an early type of perceptron with a linear transfer function. The group of 59 papers that used spectra as input can be further subdivided according to the type of spectra: 26 dealt with infrared (IR) spectra, thirteen with nuclear magnetic resonance (NMR) spectra, one with ultraviolet (UV) spectra, and seventeen with mass spectra (MS), one with IR and MS, and one used IR, MS, and NMR. Munk, Madison, and Robb [102] reported that a feed-forward net was better than linear regression for identifying functional groups present in organic compounds from their IR spectra. Borggaard and Thodberg [103] have described the OMNIS program (optimal minimal neural interpretation of spectra) wherein they use principal components analysis as a preprocessing method, and then employed cross-validation as a guide to removing network connections until a minimal network that gives good generalisation is obtained.

Three papers describe the use of neural nets for the classification and identification of crosspeaks in 2D NMR. The most recent are by Peter Johnson *et al.* [104] and by Kjaer and Poulsen.[105] Curry and Rumelhart [106] used a feed-forward, multi-layer net to classify mass spectra according to which of 100 functional groups were present. Otto and Hoerchner [107] used a hybrid of fuzzy logic and an adaptive bi-directional associative memory type of CNN, to identify UV spectra of organic compounds by finding the nearest match between the input spectrum and a set of stored spectra. Anker and Jurs [108] used a feed-forward, multi-layer net to accurately predict the ^{13}C NMR shifts of 431 keto-steroids from calculated structural descriptors. Kvasnicka [109] also described the prediction of ^{13}C NMR shifts with a multi-layer feed-forward network, but used oriented graphs as inputs. The network Kvasnicka described was structured so that it resembled the chemical structure graphs used as inputs. This is a novel approach for inputting chemical structural information, but it may be difficult to extend the procedure to the general case.

3.8 Property prediction and parameter estimation applications

Compared to the other areas of chemistry, there are few applications of neural networks for property prediction and parameter estimation. Nineteen of the 26 papers in this area were published in 1992 or 1993, indicating that this type of CNN application is being extensively investigated at present. Kito, Hattori, and Murakami [110] used a feed-forward network to predict the acid strength of mixed metal oxides. Bodor and collaborators have published two studies on the prediction of properties of organic molecules using multi-layer feed-forward networks, the first predicted the water solubility of a diverse set of compounds, [111] while the second predicted oxidation potentials of heterocyclic compounds.[112] Peterson [113] used a Counter Propagation Network (CPN) [114] to predict Kovats indices for substituted phenols with lower errors than were obtained using linear regression. The thin-layer chromatographic behaviour of 22 benzoic acid derivatives was predicted using a feed-forward neural net by Glen *et al.* [115] Egolf and Jurs [116] predicted boiling points of heterocycles using BP. Their results indicate that regression methods are still required to select the appropriate descriptors, but once the descriptors were chosen, that neural nets gave superior predictions over traditional regression techniques.

Noid, Varma-Nair, Wunderlich, and Darsey [117] inverted the usual CNN prediction process by training a multi-layer feed-forward network on heat capacities of polymers. The network was then used to estimate the two parameters of the Tarasov function, which is commonly used to predict heat capacities of polymers. The accuracy obtained using the CNN estimated parameters was significantly better than that obtained by other methods.

3.9 Reaction and reactivity prediction applications

Neural network applications for the prediction of organic reactions and reaction products are very sparse, with only eleven publications in this category to date, five of which used generalised perceptrons. Schulz and Gasteiger [118] used an associative memory method, which is a type of neural network quite different from generalised perceptrons, to predict the reactivity of single bonds in aliphatic compounds. Elrod, Maggiora, and Trenary used connectivity matrices as inputs to a generalised perceptron to predict the amounts of reaction products in electrophilic aromatic substitution (EAS) reactions as accurately as chemists and better than a rule-based expert system.[119] Further extension of their work showed that feed-forward neural nets could 'learn' simple reactivity rules for elimination, addition and cyclo-addition reactions.[120] Kvasnicka studied the same EAS reaction as Elrod *et al*, but used chemical graphs as network inputs.[121] Skelenak, Kvasnicka and Pospichal used BP to predict 1,3-dipolar cyclo-addition reactions. [122] Luce and Govind proposed a hybrid neural network-expert system for retrosynthetic analysis.[123] Marie and Villemin used a machine learning approach to extract reaction information from chemical reaction graphs.[124]

3.10 Nuclear chemistry and nuclear power applications

The 90 applications in nuclear chemistry and nuclear power plants centre on two main themes, monitoring and analysing the status of numerous sensors in nuclear reactors and detecting and identifying sub-atomic particles from decays tracks of charged particles, with most of the authors reporting the use of feed-forward networks. Guo and Uhrig [125] used a hybrid neural network which has potential applicability in other areas as well. Their hybrid net used a Kohonen SOM to cluster nuclear power plant heat output data. Then the centroids of these clusters were used to train a feed-forward net that could predict the rate of heat production with an accuracy of 0.1%. Opposing conclusions about the utility of CNNs for analysing nuclear decay events are given by Cherubini & Odorico [126] who found that a Kohonen Learning Vector Quantisation (LVQ) network performed worse than statistical methods for identifying quark decay products, while Stimpfl-Abele [127] found that feed-forward networks gave excellent results for recognising decays of charged tracks. Peterson [128] used CPN networks to classify energy levels in Curium II and Plutonium I. Cheon and Chang described the advantages of a connectionist expert system, which used a BP network, over conventional expert systems for identifying transients in nuclear power plants.

4. Neural networks: challenges and opportunities for the future

The amount of research in and applications of CNNs has increased at a phenomenal rate during the last half of the 80s. Even in chemistry, as seen in Figure 3, CNN applications have increased significantly in both variety and numbers over the last several years. Their future in chemistry appears to be bright, particularly in analytical chemistry, chemical engineering, and biomolecular informatics. However, there are a number of issues that present significant challenges to some of the other chemical applications of CNNs.[94] Two of the most important ones are how best to represent chemical information in a format suitable for neural-network applications and how to deal with the relatively meagre amount of data available on many chemical systems of interest. With regard to the first issue, a significant problem arises from the fact that input to essentially all neural networks is 'vector-like' while the most intuitive representation of chemical information is based upon structure, which can be represented as a chemical graph. To date, a general means for incorporating graph-like input into CNNs has not been developed.[109] In analytical chemistry and chemical engineering, inputs to CNNs are typically the measurable state variables of the system, which simplifies the representation problem. For biomolecular informatics applications also, the input data representation is usually straightforward, generally being the protein or nucleotide sequence itself.

With regard to the second issue, even CNNs of modest complexity can possess large numbers of weights that must be determined during network training. For example, a generalised perceptron with only two input nodes, two hidden layers of five nodes each, and a single output node possesses 51 weights. Generally, it is assumed that in order to achieve statistically reliable predictions it is necessary to have at least three samples per weights.[10,91] This would mean that for statistically reliable predictions, the simple multi-layer net described above would require a training set of about 130 samples. This represents a serious impediment to the use of CNNs in many types of problems of interest in chemistry, where the size of the dataset may be relatively small. If, for example, a QSAR study is undertaken, the amount of data needed to train a reliable CNN is substantial. Thus, the amount of synthesis needed to develop a dataset of suitable size may be such as to obviate the need for a QSAR analysis — the problem, essentially, will have already been solved. Fortunately, cases do exist in which reliable results have been achieved, even with noisy data, when the samples-to-weights ratio is less than three.[94] Nevertheless, a deeper understanding of the this issue would certainly be of great value in future work, especially in chemistry.

Even in light of the problems noted above, CNNs, due to their ability to find complex, nonlinear mappings, appear to have a future in chemistry but not, perhaps, in their 'pure' form. Recently, considerable work has been carried out on a number of hybrid systems that incorporate a variety of technologies including fuzzy logic, genetic algorithms, and computational intelligence along with neural networks. Genetic algorithms, in particular, seem to be having an impact on chemistry, as means to solve search and optimisation problems.[130,131] Although applications of hybrid systems to date are primarily in the business area, as the behaviour of these systems becomes better understood, applications in other areas including chemistry will no doubt begin to occur with increasing frequency. Recently, in fact, several papers which use hybrid neural-network methods have appeared. Otto *et al* used fuzzy logic and neural nets to design intelligent analytical instruments [76]; Guo and Uhrig used Kohonen SOMs to cluster data about nuclear power plant heat output and then trained a generalised perceptron on the cluster centroids [125]; and Mavrovouniotis and Chang described the use of hierarchies of generalised perceptrons.[132] Ham, Cohen and Cho used a hybrid net consisting of a generalised perceptron and counter-propagation network to detect biological substances in complex aqueous solutions using IR spectral data as input.[133] Otto and Hoerchner used a fuzzy neural net, where the neural-net component was an unsupervised adaptive bidirectional associative memory network, for finding the best match for the UV spectra of organic compounds.[107] D'Antone described a neural net which functioned as part of an expert system; the network was trained to recognise malfunctions in a particle detector, and the expert system suggested how to remedy the instrument failure.[134]

In addition to ways that neural networks will be directly used in solving chemical problems, they will no doubt serve chemists indirectly as well. Much new chemical information, both chemical structures, as well as reaction data and other associated data, is being generated and stored in electronic form. There is, however, a large body of chemical information that is not accessible electronically. Recently two groups of researchers have made progress towards converting printed chemical information to electronic form, and both use neural networks in the process. McDaniel and Balmuth [135] wrote the KEKULE program for OCR—Optical Chemical (Structure) Recognition and have made it commercially available. KEKULE uses a generalised perceptron neural network to decipher scanned images of chemical structures and convert them into connection tables or other easily exchanged formats. A similar, but even more ambitious project is Peter Johnson's CLiDE project at the University of Leeds.[136] The Chemical Literature Data Extraction (hence CLiDE) project is designed to take scanned images of whole pages of chemical information and extract not only the chemical structures, but also the reaction schemes, reaction arrow, reaction conditions, and auxiliary data. In addition to rule-based expert systems, Johnson's group employs neural networks for the character recognition operations in CLiDE. The result of the CLiDE program will be computer-readable chemical reaction information that could be used to build databases.

Currently, CNNs are still in the exploratory phase in chemistry, and only the first few steps along this path have been taken. Neural networks are beginning to be used as general purpose methods for classification, prediction, mapping and modelling in chemistry. Their future appears to hold great promise, and many exciting new applications of neural nets in chemistry remain to be discovered.

5. References

[1] McCulloch, W.S.; Pitts, W. A Logical Calculus of the Ideas Immanent in Nervous Activity, *Bull. of Math. Biophys.*,1943, *5*, 115-133.

[2] Hebb, D. *Organization of Behavior*, NY, John Wiley & Sons, 1949.

[3] Simpson, P.K. *Artificial Neural Systems: Foundations, Paradigms, Applications, and Implementations*; Pergamon Press: New York, 1990.

[4] Kosko, B. *Neural Networks and Fuzzy Systems: A Dynamical Systems Approach to Machine Intelligence*; Prentice Hall: Englewood Cliffs, NJ, 1992.

[5] Favlow, S.J., Ed. *Self-Organizing Methods in Modeling: GMDH Algorithms*; Marcel Dekker: New York, 1984.

[6] Rumelhart, D. E.; McClelland, J. L. *Parallel Distributed Processing, Vols. I,II*; MIT Press: Cambridge, Massachusetts, 1986.

[7] Hertz, J.; Krogh, A.; Palmer, R. G. *Introduction To The Theory of Neural Computation*; Addison-Wesley Publishing Company: Redwood City, California, 1991.

[8] Hecht-Nielsen, R. *Neurocomputing*; Addison-Wesley Publishing Company: Redwood City, California, 1990.

[9] Lippmann, R.P. An Introduction to Computing with Neural Nets; *IEEE ASSP Magazine* 1987, *April*, 4-22.

[10] White, H. Learning in Artificial Neural Networks: A Statistical Perspective; *Neural Comp.* 1989, *1*, 425-464.

[11] Simpson, P.K. *Artificial Neural Systems: Foundations, Paradigms, Applications, and Implementations*; Pergamon Press: New York, 1990, Chapter 5.

[12] Hopfield, J.J. Neural networks and physical systems with emergent collective computational abilities., *Proc. Nat. Acad. Sci.* 1982, *79*, 2554-2558.

[13] Hopfield, J.J.; Tank, D.W. Computing with neural circuits: a model, *Science* 1986, *233*, 625-633.

[14] Hirst, J. D.; Sternberg, M. J. E. Prediction of Structural and Functional Features of Protein and Nucleic Acid Sequences by Artificial Neural Networks. *Biochemistry*, 1992, *31(32)*, 7211-18.

[15] Kohonen, T. *Self-organization and Associative Memory*; Springer: Berlin, 1988.

[16] Zupan, J. 'Counter-Propagation neural networks and their applications in chemistry' Presented at 1993 Montreux International Chemical Information Conference, Oct. 19, 1993, Annecy, France.

[17] Gasteiger, J.; Zupan, J. Neural Networks in Chemistry. *Angew. Chem. Int. Ed. Eng.*, 1993, *32*, 503-527.

[18] Kolmogorov, A.N. On the Representation of Continuous Functions of Several Variables by Superposition of Continuous Functions of One Variable and Addition; *Dokl. Akad. Nauk SSSR* 1957, *114*, 953-956.

[19] Cybenko, G. Approximation by Superpositions of a Sigmoidal Function; *Math. Contr. Sig. Sys.* 1989, *4*, 303-314.

[20] Hecht-Nielsen, R. Theory of the Backpropagation Neural Network. In *Proc. Int. Joint Conf. Neural Networks, Vol. I*; IEEE Press: New York, 1989.

[21] Stinchcombe, M.; White, H. Universal Approximation Using Feedforward Networks with Non-Sigmoid Hidden Layer Activation Functions. In *Proceedings of the International Joint Conference on Neural Networks (Washington, D.C., June 1989); 1989; Vol.I, pp. 607-611.*

[22] Chester, D.L. Why Two Hidden Layers are Better Than One. In *Proceedings of the International Joint Conference on Neural Networks (Washington, D.C., January 1989); 1989; Vol.I, pp. 265-268.*

[23] Minsky, M.; Papert, S. *Perceptrons*; MIT Press: Cambridge, MA, 1969.

[24] Rumelhart, D.E.; Hinton, G.E.; Williams, R.J. Learning Representations by Back-Propagating Errors, *Nature* 1986, *323*, 533-536.

[25] Werbos, P. Beyond Regression: New Tools for Prediction and Analysis in the Behavioral Sciences, Ph.D. dissertation, Harvard University, Cambridge, MA, 1974.

[26] Hertz, J.; Krogh, A.; Palmer, R. G. *Introduction To The Theory of Neural Computation*; Addison-Wesley Publishing Co: Redwood City, CA, 1991, p 102-104.

[27] Klimasauskas, C. C. Neural Networks: A Short Course. From Theory to Application. *PC AI*, 1988, *4*, 26-30.

[28] Denning, P. J. Neural Networks. *Am. Scientist*, 1992, *80*, 426-9.

[29] Lapedes, A.; Farber, R. How Neural Networks Work. In *Neural Information Processing Systems*, Denver 1987; Anderson, D.Z., Ed.; American Instit. of Physics: New York, 1988; 442-456.

[30] CA File database, Chemical Abstracts Service, American Chemical Society, Columbus, OH, 1967-1993.

[31] INSPEC Database, IEEE, available on STN International, Chemical Abstracts Service, American Chemical Society, Columbus, OH, 1969-1993.

[32] COMPUSCIENCE Database, FIZ Karlsruhe, Germany, available on STN International, Chemical Abstracts Service, American Chemical Society, Columbus, OH, 1982-1993.

[33] DISSABS Database, UMI, Inc, Ann Arbor, MI, available on STN International, Chemical Abstracts Service, American Chemical Society, Columbus, OH, 1861-1993.

[34] Jurs, P. C.; Kowalski, B. R.; Isenhour, Thomas L.; Reilley, C. N. Computerized Learning Machines Applied to Chemical Problems.

Convergence Rate and Predictive Ability of Adaptive Binary Pattern Classifiers. *Anal. Chem.*, 1969, **41***(6)*, 690-5.

[35] Isenhour, T. L.; Jurs, P. C. Learning Machines. *Comput. Chem. Instrum.*, 1973, **1**, 285-330.

[36] Jurs, P. C. Binomial Distribution Statistics Applied to Minimizing Activation Analysis Counting Errors. An Analog Computer Controlled Gamma-Ray Spectrometer for Comparative Activation Analysis. Computerized Learning Machines for the Interpretation of Low Resolution Mass Spectrometry Data. 1969, Ph.D. dissertation, University of Washington, Seattle, WA, 95 pp.

[37] Roat, S. D. The Application of Neural Networks and System Cultivation with a Nonlinear Optimal Control Algorithm for the Chemical Process Industry. 1991, Ph.D. dissertation, Univ. Tennessee, 261 pp.

[38] Haesloop, D. G. System Identification and Control using Neural Networks with Combined Linear and Non-Linear Mapping Functionality. 1991, Ph.D. dissertation, Univ. Washington, 184 pp.

[39] Bhat, N. D. System Identification and Control of Dynamic Chemical Processes Using Backpropagation Networks. 1991, Ph.D. dissertation, Univ. Maryland, 262 pp.

[40] Jurs, P.C. ACS National Meeting, Chicago, August 22-26, 1993. Computers in Chemistry Division Abstracts no. COMP34, COMP87,and COMP102.

[41] Zupan, J.; Gasteiger, J. *Neural Networks for Chemists: A Textbook*; VCH: Weinheim, Germany, 1993.

[42] Elrod, D. W.; Maggiora, G. M. Artificial Neural Networks: A New Computational Paradigm with Applications in Chemistry. In *Online Information 92*, 16th International Online Information Meeting Proceedings, London, UK, Dec. 8, 1992, Learned Information: Oxford, 109-125.

[43] Lacy, M. E. Neural Network Technology and its Application in Chemical Research. *Tetrahedron Comput. Methodol.*, 1990, **3***(3-4)*, 119-28.

[44] Schmuller, J. Neural Networks and Environmental Applications. In *ACS Symp. Ser. 431, Expert Systems for Environmental Applications*; Hushon, J., Ed; American Chemical Society: Washington, D.C., 1990; 52-68.

[45] Wythoff, Barry J. Backpropagation neural networks. A tutorial. *Chemom. Intell. Lab. Syst.* 1993, **18***(2)*, 115-55.

[46] Holcomb, T. R.; Morari, M. PLS/Neural Networks. *Comput. Chem. Eng.*, 1992, **16***(4)*, 393-411.

[47] Qin, S. J.; McAvoy, T. J. Nonlinear PLS Modeling Using Neural Networks. *Comput. Chem. Eng.*, 1992, **16***(4)*, 379-91.

[48] Kramer, M. A. Nonlinear Principal Component Analysis Using Autoassociative Neural Networks. *AIChE J.*, 1991, **37***(2)*, 233-43.

[49] De Veaux, R. D.; Psichogios, D. C.; Ungar, L. H. A comparison of two nonparmatric estimation schemes: MARS and neural networks. *Comput. Chem. Eng.* 1993, **17***(8)*, 819-837.

[50] de B. Harrington, Peter; Pack, Brian W. FLIN: fuzzy linear interpolating network. *Anal. Chim. Acta*, 1993 **277***(2)*, 189-197.

[51] Bulsari, A. B.; Kraslawski, A.; Saxen, H. Implementing a fuzzy expert system in an artificial neural network. *Comput. Chem. Eng., 17(Suppl., European Symposium on Computer Aided Process Engineering-2, 1992)* 1993, S405-S410.

[52] Kateman, G.; Smits, J. R. M. Colored information from a black box? Validation and evaluation of neural networks. *Anal. Chim. Acta* 1993, **277**(2), 179-88.

[53] Livingstone, D. J.; Hesketh, G.; Clayworth, D. Novel Method for the Display of Multivariate Data Using Neural Networks. *J. Mol. Graphics*, 1991, **9***(2)*, 115-18.

[54] Reibnegger, G.; Weiss, G.; Wachter, H. Self-organizing neural networks as a means of cluster analysis in clinical chemistry. *Eur. J. Clin. Chem. Clin. Biochem.* 1993, **31***(5)*, 311-16.

[55] Androsiuk, J.; Kulak, L.; Sienicki, K. Neural network solution of the Schroedinger equation for a two-dimensional harmonic oscillator. *Chem. Phys.* 1993, **173***(3)*, 377-83.

[56] Gingrich, Chester G.; Kuespert, Daniel R.; McAvoy, Thomas J. Modeling human operators using neural networks. *Adv. Instrum. Control* 1990, **45***(3)*, 1481-91.

[57] Kennedy, Max J.; Prapulla, S. G.; Thakur, M. S. Designing fermentation media: a comparison of neural networks to factorial design. *Biotechnol. Tech.* 1992, **6***(4)*, 293-8.

[58] Hattori, Tadashi; Kito, Shigeharu Catalyst design: artificial intelligence and neural networks. *Petrotech (Tokyo)* 1992, **15***(10)*, 917-21.

[59] Lelu, A. From data analysis to neural networks. New prospects for efficient browsing through databases. *J. Inf. Sci.: Principles and practice.* 1991, **17**(1), 1-12.

[60] Hoskins, J. C.; Himmelblau, D. M. Artificial Neural Network Models of Knowledge Representation in Chemical Engineering. *Comput. Chem. Eng.*, 1988, **12***(9-10)*, 881-90.

[61] *Computers in Chemical Engineering*, 1992, **16***(4)*.

[62] Bugmann, G.; Lister, J. B.; Von Stockar, U. Characterizing Bubbles in Bioreactors Using Light or Ultrasound Probes: Data Analysis by Classical Means and by Neural Networks. *Can. J. Chem. Eng.*, 1991, **69***(2)*, 474-80.

[63] Venkatasubramanian, V.; Chan, K. A Neural Network Methodology for Process Fault Diagnosis. *AIChE J.*, 1989, **35***(12)*, 1993-2002.

[64] Joseph, B.; Wang, F. H.; Shieh, D. S. S. Exploratory Data Analysis: a Comparison of Statistical Methods with Artificial Neural Networks. *Comput. Chem. Eng.*, 1992, **16***(4)*, 413-23.

[65] Bos, Martinus; Bos, Albert; van der Linden, Willem E. Data processing by neural networks in quantitative chemical analysis. *Analyst* (Cambridge, UK) 1993, **118***(4)*, 323-328.

[66] Chang, S. M.; Iwasaki, Y.; Suzuki, M.; Tamiya, E.; Karube, I.; Muramatsu, H. Detection of Odorants Using an Array of Piezo-electric Crystals and Neural Network Pattern Recognition. *Anal. Chim. Acta*, 1991, **249***(2)*, 323-9.

[67] Nakamoto, T.; Fukuda, A.; Moriizumi, T. Perfume and flavor identification by odor-sensing system using quartz-resonator sensor array and neural network pattern recognition. *Sens. Actuators, B* 1993, *B10(2)*, 85-90.

[68] Akay, M. Noninvasive diagnosis of coronary artery disease using a neural network algorithm. *Biological Cybernetics* 1992, **67***(4)*, 361-367.

[69] Rogers, S. J.; Fang, J. H.; Karr, C. L.; Stanley, D. A. Determination of Lithology From Well Logs Using a Neural Network. *AAPG Bull.*, 1992, **76**(5), 731-9.

[70] Bos, A.; Bos, M.; Van der Linden, W. E. Artificial Neural Networks as a Tool for Soft-Modeling in Quantitative Analytical Chemistry: the Prediction of the Water Content of Cheese. *Anal. Chim. Acta*, 1992, **256***(1)*, 133-44.

[71] Long, J. R.; Mayfield, H. T.; Henley, M. V.; Kromann, P. R. Pattern Recognition of Jet Fuel Chromatographic Data by Artificial Neural Networks With Back-Propagation of Error. *Anal. Chem.*, 1991, **63***(13)*, 1256-61.

[72] Bevan, A. V.; Bryant, S. A.; Clark, J. M.; Reid, K. The application of neural networks to particle shape classification. *J. Aerosol Sci.* 1992, **23***(Suppl. 1)*, S329-S332.

[73] Zupan, J. Can an Instrument Learn From Experiments Done by Itself? *Anal. Chim. Acta*, 1990, **235***(1)*, 53-63.

[74] Freeman, R. High resolution NMR using selective excitation. *J. Molec. Struct.* 1992, **266,** 39-51.

[75] Casale, J. F.; Watterson, J. W. A computerized neural network method for pattern recognition of cocaine signatures. *J. Forensic Sci.* 1993, **38***(2)*, 292-301.

[76] Otto, M.; George, T.; Schierle, C.; Wegscheider, W. Fuzzy logic and Neural Networks — Applications to Analytical Chemistry. *Pure Appl. Chem.*, 1992, **64***(4)*, 497-502.

[77] Stolorz, P.; Lapedes, A.; Xia, Y. Predicting Protein Secondary Structure Using Neural Net and Statistical Methods. *J. Mol. Biol.*, 1992, **225***(2)*, 363-77.

[78] Muskal, S. M.; Kim, S. H. Predicting Protein Secondary Structure Content. A Tandem Neural Network Approach. *J. Mol. Biol.*, 1992, **225***(3)*, 713-27.

[79] Friedrichs, M. S.; Goldstein, R. A.; Wolynes, P. G. Generalized Protein Tertiary Structure Recognition Using Associative Memory Hamiltonians. *J. Mol. Biol.*, 1991, **222***(4)*, 1013-34.

[80] Ferran, E.A.; Ferrara, P. Topological Maps of Protein Sequences. *Biological Cybernetics*, 1991, **65(6)**, 451-8.

[81] Sumpter, B. G.; Noid, D. W. Potential Energy Surfaces for Macromolecules. A Neural Network Technique. *Chem. Phys. Lett.*, 1992, **192(5-6)**, 455-462.

[82] Bohm, G. Protein Folding and Deterministic Chaos: Limits of Protein Folding Simulations and Calculations. *Chaos, Solitons and Fractals*, 1991, **1(4)**, 375-82.

[83] Bohr, Jakob; Bohr, Henrik; Brunak, Soren; Cotterill, Rodney M. J.; Fredholm, Henrik; Lautrup, Benny; Petersen, Steffen B. Protein structures from distance inequalities. *J. Mol. Biol.* 1993, **231(3)**, 861-869.

[84] Uberbacher, E. C.; Mural, R. J. Locating Protein-Coding Regions in Human DNA Sequences by a Multiple Sensor-Neural Network Approach. *Proc. Natl. Acad. Sci. USA.*, 1991, **88(24)**, 11261-5.

[85] Stormo, G. D.; Schneider, T. D.; Gold, L.; Ehrenfeucht, A. Use of the Perceptron Algorithm to Distinguish Translational Initiation Sites in *E. coli. Nucleic Acids Res.*, 1982, **10(9)**, 2997-3011.

[86] Arrigo, P.; Giuliano, F.; Scalia, F.; Rapallo, A.; Damiani, G. Identification of a New Motif on Nucleic Acid Sequence Data Using Kohonen's Self-Organizing Map. *Comput. Appl. Biosci.*, 1991, **7(3)**, 353-7.

[87] Ladunga, I.; Czako, F.; Csabai, I.; Geszti, T. Improving Signal Peptide Prediction Accuracy by Simulated Neural Network. *Comput. Appl. Biosci.*, 1991, **7(4)**, 485-7.

[88] Claverie, J. M.; Sauvaget, I. WOBB.C: A Portable Software Package for Defining and Searching Ambiguous Sequence Patterns. *Protein Sequences Data Anal.*, 1991, **4(2)**, 119-21.

[89] Frishman, D.; Argos, P. Recognition of distantly related protein sequences using conserved motifs and neural networks. *J. Mol. Biol.* 1992, **228(3)**, 951-962.

[90] Wikel, James H.; Dow, Ernst R. The use of neural networks for variable selection in QSAR. *Bioorg. Med. Chem. Lett.* 1993, **3(4)**, 645-51.

[91] Andrea, T. A.; Kalayeh, H. Applications of Neural Networks in Quantitative Structure-Activity Relationships of Dihydrofolate Reductase Inhibitors. *J. Med. Chem.* 1991, **34**, 2824-2836.

[92] Aoyama, T.; Ichikawa, H. Reconstruction of Weight Matrices in Neural Networks—a Method of Correlating Outputs With Inputs. *Chem. Pharm. Bull.* 1991, **39(5)**, 1222-1228.

[93] Tetko, I. V.; Luik, A. I.; Poda, G. I. Applications of neural networks in structure-activity relationships of a small number of molecules. *J. Med. Chem.* 1993, **36(7)**, 811-14.

[94] Maggiora, G. M.; Elrod, D. W.; Trenary, R. G. Computational Neural Nets as Model-Free Mapping Devices. *J. Chem. Inf. Comput. Sci.* 1992, **32**, 732-741.

[95] Livingstone, David J.; Manallack, David T. Statistics using neural networks: chance effects. *J. Med. Chem.* 1993, **36***(9)*, 1295-1297.

[96] Zitko, V. Prediction of Biodegradability of Organic Chemicals by an Artificial Neural Network. *Chemosphere*, 1991, **23***(3)*, 305-12.

[97] Chastrette, M.; De Saint Laumer, J. Y. Structure-Odor Relation-ships Using Neural Nets. *Eur. J. Med. Chem.*, 1991, **26***(8)*, 829-33.

[98] Hussain, A. S.; Yu, X.; Johnson, R. D. Application of Neural Computing in Pharmaceutical Product Development. *Pharm. Res.*, 1991, **8***(10)*, 1248-52.

[99] Cambon, Benoit; Devillers, James New trends in structure-biodegradability relationships. *Quant. Struct.-Act. Relat.* 1993, **12***(1)*, 49-56.

[100] Weinstein, J. N.; Kohn, K. W.; Grever, M. R.; Viswanadhan, V. N.; Rubinstein, L. V.; Monks, A. P.; Scudiero, D. A.; Welch, L.; Koutsoukos, A. D.; *et al.* Neural computing in cancer drug development: predicting mechanism of action. *Science (Washington, D. C.)* 1992, **258***(5081)*, 447-51.

[101] Rose, V. S.; Croall, I. F.; MacFie, H. J. H. An Application of Unsupervised Neural Network Methodology (Kohonen Topology- Preserving Mapping) to QSAR Analysis. *Quant. Struct.-Act. Relat.*, 1991, **10***(1)*, 6-15.

[102] Munk, M. E.; Madison, M. S.; Robb, E. W. Neural Network Models for Infrared Spectrum Interpretation. *Mikrochim. Acta*, 1991, **2***(1-6)*, 505-14.

[103] Borggaard, C.; Thodberg, H. H. Optimal Minimal Neural Interpretation of Spectra. *Anal. Chem.*, 1992, **64***(5)*, 545-51.

[104] Corne, Simon A.; Fisher, Julie; Johnson, A. Peter; Newell, William R. Cross-peak classification in two-dimensional nuclear magnetic resonance spectra using a two-layer neural network. *Anal. Chim. Acta* 1993, **278***(1)*, 149-58.

[105] Kjaer, M.; Poulsen, F. M. Identification of 2D Proton NMR Antiphase Cross Peaks Using a Neural Network. *J. Magn. Reson.*, 1991, **94***(3)*, 659-63.

[106] Curry, B.; Rumelhart, D. E. MSnet: A Neural Network Which Classifies Mass Spectra. *Tetrahedron Comput. Methodol.*,1990, **3***(3-4)*, 213-37.

[107] Otto, M.; Hoerchner, U. (1990) Application of Fuzzy Neural Network to Spectrum Identification, *Software Dev. Chem. 4, Proc. Workshop Comput. Chem.*, Meeting Date 1989, Ed: Gasteiger, J.; Springer: Berlin, 1990; p 377-84.

[108] Anker, L. S.; Jurs, P. C. Prediction of Carbon-13 Nuclear Magnetic Resonance Chemical Shifts by Artificial Neural Networks. *Anal. Chem.*, 1992, **64***(10)*, 1157-64.

[109] Kvasnicka, V. An Application of Neural Networks in Chemistry. Prediction of ^{13}C NMR Chemical Shifts. *J. Math. Chem.* 1991, **6**, 63-76.

[110] Kito, S.; Hattori, T.; Murakami, Y. Estimation of the Acid Strength of Mixed Oxides by a Neural Network. *Ind. Eng. Chem. Res.*, 1992, **31***(3)*, 979-81.

[111] Bodor, N.; Harget, A.; Huang, M. J. Neural Network Studies. 1. Estimation of the Aqueous Solubility of Organic Compounds. *J. Am. Chem. Soc.*, 1991, **113***(25)*, 9480-3.

[112] Brewster, M. E.; Huang, M. J.; Harget, A.; Bodor, N. Reactivity of Biologically Important Reduced Pyridines. 10. Neural Network Studies. 2. Use of a Neural Net to Estimate Oxidation Energies for Substituted Dihydropyridines and Related Heterocycles. *Tetrahedron*, 1992, **48***(17)*, 3463-72.

[113] Peterson, K. L. Counter-Propagation Neural Networks in the Modeling and Prediction of Kovats Indexes for Substituted Phenols. *Anal. Chem.*, 1992, **64***(4)*, 379-86.

[114] Hecht-Nielsen, R. *Neurocomputing*; Addison-Wesley Publishing Company: Redwood City, CA, 1990, p 147-153.

[115] Glen, R. C.; Rose, V. S.; Lindon, J. C.; Ruane, R. J.; Wilson, I. D.; Nicholson, J. K. Quantitative Structure-Chromatography Relationships: Prediction of TLC behavior Using Theoretically Derived Molecular Properties. *J. Planar Chromatogr.-Mod. TLC*, 1991, **4,** 432-8.

[116] Egolf, Leanne M.; Jurs, Peter C. Prediction of boiling points of organic heterocyclic compounds using regression and neural network techniques. *J. Chem. Inf. Comput. Sci.* 1993, **33,** 616-625.

[117] Noid, D. W.; Varma-Nair, M.; Wunderlich, B.; Darsey, J. A. Neural Network Inversion of the Tarasov Function Used for the Computation of Polymer Heat Capacities, *J. Therm. Anal.*, 1991, **37***(10)*, 2295-300.

[118] Schulz, Klaus Peter; Gasteiger, Johann Elucidation of chemical reactivity using an associative memory system. *J. Chem. Inf. Comput. Sci.* 1993, **33***(3)*, 395-406.

[119] Elrod, D. W.; Maggiora, G. M.; Trenary, R. G. Applications of Neural Networks in Chemistry. 1. Prediction of Electrophilic Aromatic Substitution Reactions. *J. Chem. Inf. Comput. Sci.* 1990, **30***(4)*, 477-84.

[120] Elrod, D. W.; Maggiora, G. M.; Trenary, R. G. Applications for Neural Networks in Chemistry. 2. A General Connectivity Representation for the Prediction of Regiochemistry. *Tetrahedron Comput. Methodol.* 1990, **3**, 163-74.

[121] Kvasnicka, V.; Pospichal, J. Application of Neural Networks in Chemistry. Prediction of Product Distribution of Nitration in a Series of Monosubstituted Benzenes. *THEOCHEM* 1991, **81***(3-4)*, 227-42.

[122] Sklenak, S.; Kvasnicka, V.; Pospichal, J. Prediction of regioselectivity of 1,3-dipolar cycloaddition reactions by neural networks. *Acta Chim. Hung.* 1993, **130***(1)*, 103-110.

[123] Luce, H. H.; Govind, R. Neural Network Applications in Synthetic Organic Chemistry. I. A Hybrid System Which Performs Retrosynthetic Analysis. *Tetrahedron Comput. Methodol.* 1990, **3***(3-4)*, 143-61.

[124] Marie, S.; Nicolle-Adam. A.; Villemin, D. MAROCO: A Learning Machine on Object-Based Representations in Organic Chemistry. in

Proceedings of the First International Conference on Knowledge Modeling and Expertise Transfer, 22-24 April 1991, Sophia-Antipolis, France; IOS: Amsterdam, 1991, p 115-29.

[125] Guo, Z.; Uhrig, R. E. Use of Artificial Neural Networks to Analyse Nuclear Power Plant Performance. *Nucl. Technol.*, 1992, **99(1)**, 36-42.

[126] Cherubini, A.; Odorico, R. Discrimination of p.hivin. p.fwdarw. t.mchlt.t Events by a Neural Network Classifier. *Z. Phys. C: Part. Fields*, 1992, **53(1)**, 139-48.

[127] Stimpfl-Abele, G. Recognition of Decays of Charged Tracks With Neural Network Techniques. *Comput. Phys. Commun.*, 1991, **67(2)**, 183-92.

[128] Peterson, K. L. Classification of Curium II and Plutonium I Energy Levels Using Counterpropagation Neural Networks. *Phys. Rev. A*, 1991, **44(1)**, 126-38.

[129] Klimasauskas, C. C. Hybrid Technologies: More Power for the Future. In *Advanced Technology for Developers*, **1**, August 1992, High-Tech Communications: Sewickley, PA, 1992, 17-20.

[130] Lucasius, C. B.; Kateman, G. Understanding and using genetic algorithms. Part 1. Concepts, properties and context. *Chemom. Intell. Lab. Syst.* 1993, **19(1)**, 1-33.

[131] Hibbert, D. B. Genetic algorithms in chemistry. *Chemom. Intell. Lab. Syst.* 1993, **19**, 277-293.

[132] Mavrovouniotis, M. L.; Chang, S. Hierarchical Neural Networks, *Comput. Chem. Eng.*, 1992, **16(4)**, 347-69.

[133] Ham, F.M.; Cohen, G.M.; Cho, B. Improved Detection of Biological Substances Using a Hybrid Neural Network and Infrared Absorption Spectroscopy. In *Conference Proceedings: IJCNN-91-Seattle: International Joint Conference on Neural Networks*, 1991, **1**, 227-32.

[134] D'Antone, I. A Neural Network in an Expert Diagnostic System. *IEE. Trans. Nucl. Sci.*, 1992, **39(2, Pt. 1)**, 58-62.

[135] McDaniel, J.R.; Balmuth, J.R. Kekule: OCR-Optical Chemical (Structure) Recognition. *J. Chem. Inf. Comput. Sci.* 1992, **32**, 373-378.

[136] Ibison, P,; Jacquot, M.; Kam, F.; Neville, A.G.; Simpson, R.W.; Tonnelier, C.; Venczel, T.; Johnson, S.P. Chemical Literature Data Extraction: the CLiDE Project. *J. Chem. Inf. Comput. Sci.* 1993, **33**, 338-344.

Changes in the demand for information in the chemical industry

Wolfgang T. Donner

Zentrale Dienste Forschung, Bayer AG, 5090 Leverkusen, Germany

Several separate events happened recently which seem to indicate a change in the way the chemical industry uses and deals with information. From our company point of view, I would mention the following:

- The IDC (Internationale Dokumentationsgesellschaft für Chemie) has ceased its activities

- Bayer published for many decades the *Textil-Bericht* (Textile Report) and the *Hochmolekular-Bericht* (Macromolecular Report), containing abstracts of publications and patents in the fields of dye-stuffs and polymers, respectively. Both reports discontinued their appearance in 1992.

- Many companies no longer obtain patent information in printed form but on CD-ROM.

These events can be considered symptomatic of a general development which can be traced back to fundamental changes. It seems inappropriate to view these as the results of one single factor. To assume that these events are merely the result of the current economic recession seems to be inadequate. As reasons for this development, one will find the following factors:

(1) Changes that are conditioned by the development of computer technology

(2) Changes on the information market

(3) Production costs

(4) Changes in the demand for information in the chemical industry.

This paper aims to start the discussion on a revolution that obviously is underway rather than to present ready-made solutions. This revolution concerns information management and dissemination.

1. Computer technology

Obviously, it needs no further explanation that the information market of today is governed, above all, by the development of computer technology. Within the last five decades, computing power, as measured by computing speed, has increased by roughly two orders of magnitude per decade.

This increase has not only had the consequence that quantum chemistry can be employed to calculate ever larger molecules with higher accuracy. The increase in performance has also been used to improve the intelligence and hence user-friendliness of programs. For instance, 25 years ago it was necessary to employ specialists to encode chemical information. The development of techniques to communicate with the computer by drawing structural diagrams on a screen became possible as soon as programs were available to analyse the graph automatically and to store chemical information in a searchable way.

Storage capacity has also increased in the same way. In the fifties, drum storage was — from today's perspective — ridiculously small. It was superseded by the magnetic tape, which allowed only very slow and sequential access. Magnetic disk technology offered a much faster and larger storage medium. Further dimensions are opened today by optical storage media.

As a consequence of this development, an increasing quantity of data could be stored and processed in an ever-shorter time. Parallel to this, the user-friendliness of programs was improved. Today, the end-user himself can perform searches with current information systems. This, of course, changes the situation of the information specialist in the chemical industry. Progress is made obvious by comparing serial retrieval systems based on a fragment code, such as GREMAS, with online systems based on topology, such as CAS Online, Beilstein Online and especially the in-house systems such as DARC, MACCS, ORAC, RESY, etc.

One also has to consider the fact that today the majority of publications are produced on a computer. This too will have several consequences for the market for information.

2. Changes to the information market

The above-mentioned products — the GREMAS database, Hochmolekular-Bericht, Textil-Bericht — were, for a long time, essential products. Within the last decade, however, other products such as CAS Online or the Derwent files have entered this market. Initially, the coding and retrieval techniques offered by the newcomers were, in many cases, insufficient. However, we are observing a constant improvement in these services. They have not remained alone: other files have joined the fray in the same market.

These developments have been partly triggered by the development of computer techniques. However, when an electronic database is built up in addition to a printed product, it presents a severe problem for every information provider. Very often, external competition and the need for increased timeliness force this duplicate strategy. At the same time, higher production costs and internal competition between two product forms, printed and electronic version, are the consequences.

Again, one should remember that the majority of scientific publications are produced on a computer. The diskette and the printed versions go to the publisher who produces the journal or book from them. Besides the distribution of this printed form of information, either the publisher or organisations such as Chemical Abstracts Services produce databases in parallel to these, as another medium for distribution. Meanwhile, we obtain information via the following media:

- Print
- Online databases
- Magnetic tape
- Diskettes
- Optical storage (e.g. CD-ROM).

In some cases the printed medium is merely necessary to justify the publisher's expenses. The function of the publisher is to make information generally accessible and quotable. This function is traditionally bound to the print medium. The user, however, will increasingly prefer electronic media.

An article appeared recently in *Chemical & Engineering News* covering the questions of online journals, electronic publishing and on-demand printing. Here, the statement was made that we have to expect a publishing revolution comparable to that of the invention of Johannes Gutenberg.

Seen from a pessimistic viewpoint, each publisher is forced to enter ever deeper into this morass of problems, as the restriction of any service to a printed product alone often is tantamount to surrender.

A solution to the economic question is certainly decisive for the survival of any service. It would be naive to expect that the market will accept any price. Even the best marketing cannot help if similar information at a significantly lower price exists. Therefore, besides the quality of the information, costs are becoming even more important.

3. Production Costs

The classical form of production of an information service is outlined by the following scheme:

Selection of literature

Abstracting of literature

Input in card-files / printed product / database / CD-ROM

70–80% of the costs for such products are personnel costs. A very thorough and detailed analysis and control of these costs is thus absolutely essential. Here one has to consider questions of how modern technology can be used to reduce the required manpower or how to move away from a high salary environment. Those who find here the more efficient solutions will survive.

Another phenomenon has also to be considered: the question of the value of information. One cannot deny that, in addition to the classical triad of production factors — manpower, raw materials and capital — one has to consider information.

Contrary to the classical factors, there exist up until now only rough ideas as to how to evaluate the market value of information. Even if there are some differences between classical production factors and information, I have the impression that we are currently living in a phase where the market value of information has still to be determined. A few examples from our own location illustrate this.

For our printed products Hochmolekular-Bericht and Textil-Bericht, we found that the majority of information processed here was also processed at other places (e.g.

Derwent, KKF, CAS, Rapra, etc.). The following factors spoke for an additional production step by experts in our company:

- Higher selectivity
- Better lay-out of the printed product
- Familiarity.

These are essentially quality arguments.

The strongest argument against internal production came from considering the high costs. In principle, we encounter here the old argument of price versus higher quality. The result of an internal analysis was that it is possible to take the existing information from the market, and to improve that mainly automatically, to attain a quality comparable to that of internal production, at nearly half the price. Very often, insufficient retrieval techniques offered in the market are the reason for lesser quality.

The decision regarding the IDC question can be understood in a similar way. The quality of the IDC databases was never questioned; primarily economic aspects led to the decision. One may regret that in the past not enough attention was paid to the changes within the information market. A decreasing quality advantage is related to disproportionately increasing costs. It was certainly tried too late, by the use of new techniques, to allow access by the end-user to the databases and to reduce the costs, either by reducing the effort for the input, by enlarging the user community, or both. Of course, there were also other aspects that forced this decision, as we shall see later.

This development is difficult to understand for those producing information at a high quality level to fulfil the exacting requirements of the information specialists in the chemical industry. Not that the information specialists are lowering their demands on quality. The economic situation of the chemical industry forces its members to look twice at every penny before it is spent.

From the publisher's point of view, information distribution in printed form can rely on a certain common understanding between producer and user on the value of the product: The more pages, the higher the price. This fundamental agreement rests on the unrealistic assumption that everybody — at least in principle — is willing to read the book from the first page to the last. Electronic publishing offers new aspects: the computer searches for exactly the information I am interested in. Having this in mind, we immediately see the financial problem yet unsolved: what is the exact value of the publisher's activity?

4. Demand for information in industry

We also find changes which result from the fact that information is becoming more and more an additional production factor. Here, it is essential that information be held in readiness in a timely and cost-efficient manner.

With regard to costs, we observe in our company, as well as in others, a drastic change. In the past, information was considered as part of the company infrastructure. The information department thus lived from a budget fixed by the management. The receiver of information, however, is, in most cases, not the management but the bench chemist, as the research chemist generally has the highest demand for information. This separation between those who have the demand and those who decide on the budget was called more and more into question, for several reasons. Today there is a strong trend towards removing this separation by charging the chemist according to the service he obtains. There is a recognisable movement towards a market-oriented performance of the information departments.

One understands immediately that this also had a serious consequence for IDC. The relationship between IDC and its partner companies was ruled by the former model, where information was considered as part of the infrastructure within the companies. The costs for IDC were not allocated to the companies according to actual use, but rather to the *potential* use by each company. As a measure for this, the number of chemists within each company was used.

With the change from the infrastructure model to a market-oriented model (payment by use) within the companies, IDC found itself with an insoluble problem.

Apart from financial aspects, timeliness becomes an ever more important factor. The time factor expresses itself in the fact that we find today in each company that all available technical solutions are used to give the chemist himself direct access to information. One will find in-house systems in any company. In our company, for example, the chemists in the research departments of many divisions have access to the internal network. This development will certainly continue. Today about 400–500 chemists in our company are users of the internal system, which collects and processes information relevant to them in specific files.

Naturally, for them the computer is just a tool, not an end in itself. They do not have the time to become familiar with the technical differences of all the external systems which have information relevant to them. It is necessary for the chemist to access external databases with the same familiar system he uses for internal information. There is obviously a market here for external database providers.

To utilise this potential, it seems necessary to develop interfaces between internal systems and external databases, which allow access to external databases from within the internal system. Ideally, external files should be as easily accessible from the internal system as internal files.

That does not necessarily mean a downloading of external files (or parts of them) into internal databases. Open interfaces can be another solution to reach that goal. Again, one has not only to consider technical aspects, but also economic ones. The information specialists in industry must co-operate with the specialists on the side of the host to create the necessary technical provisions.

From our current experience, the following approach would seem to be suitable: there is no sense in downloading completely extensive public files into an in-house system. In particular, the one-off costs for the downloading and the continual effort of keeping them up-to-date seem to be too high to justify this.

The situation might be different for a reduced set of external information covering just a specific topic of current relevance in research. Here, there certainly exists a great demand to combine internal research results with external information and to build up a project-oriented database which is only of passing interest. In principle, such a specific database can be compared with the card file, which most research chemists have for their fields of interest. In these files, the chemist collects and orders the information, according to aspects that best reflect his point of view.

5. Conclusion

Both sides, publisher and user in the chemical industry, are forced to react to new publishing technology by developing new ways of dissemination and use of chemical information. Both sides certainly have only vague ideas concerning the impact of new technology. Finally, from the publisher's side also, severe economic risks are involved in that development. Both sides have to adapt to a new pricing policy, since the traditional ways such as counting pages or connect hours to a database, will be inadequate.

On the other hand, any publisher can count on the fact that the chemical industry will need information as long as there is any research. It might be tempting to demand lower prices for the information, but industry has to be aware of the fact that an information market exists only if prices allow its survival.

Any reasonable solution has to consider the different situations of the parties involved:

- Information producer / provider in the market (publisher)
- Information department in industry
- End-user of information in industry.

In particular, for those who are abstracting or critically evaluating published information for the market, the development outlined here will have certain consequences. In Germany, we traditionally find a very high standard in this field, where, for instance, names such as Beilstein, ChemInform, Gmelin, Houben-Weyl and others can be mentioned. They rely on the conditions of a specific market which they support. It seems that these conditions are changing. If the established institutions in this area do not wish to fail due to economic problems, they will have to react to these changes. Their main asset is a high standard of quality, but economic constraints become ever more important.

Perhaps a solution will be found by a stronger, synergistic co-operation between these different institutions. Any co-operation is, however, only advantageous if it results in a significant reduction of production costs. It is certainly not easy to proceed along that path, as not only is a common understanding among different institutions required, but they have, at the same time, to give up, at least partially, their own traditions and the internal identification for a particular product. To find a new way to success, a great deal of imagination and entrepreneurial skill will be required.

As all these changes are happening at the same time, the dialogue between both parties — publisher, including database producer, and the chemical industry — will help both sides to find a well balanced way from the past into an unknown but certainly different future.

Acknowledgements

The author has to thank many colleagues and friends for fruitful discussions. Especially the ideas and suggestions of Dr. B. Dunne, Dr. Michels, Prof. R. Luckenbach, helped considerably while preparing this paper.

The survival of the Technical Information Centre in the 1990s: deductions from a recent survey

Monica Pronin

Director, Central Abstracting & Information Services, American Petroleum Institute, New York City, USA

Introduction

The Technical Information Centre (TIC) plays a vital role in keeping its company competitive. Its many services invariably include literature searching, document delivery, referral and library maintenance, and may often also include records management, end user training and information analysis.

The result is cost avoidance. This is achieved by the TIC through its services, which provide current awareness and problem solving in support of research, operations, patenting, litigation and strategic or economic planning. The most direct benefit is the maximization of the effectiveness of research, including the non-repetition of work already done elsewhere. There is no question: this affects the bottom line. However, TICs may provide less directly even greater benefits to operations by, for example, helping engineers to solve technical problems faster; to keep abreast of new equipment, processes and products; to stay aware of environmental concerns and solutions; to meet health, safety and quality standards; and to develop new standards. This too affects the bottom line.

The value of the TIC is evident, or it should be. In the best of times, however, the TIC is often not fully appreciated in many companies. Its ultimate achievement — cost avoidance — is hard to track and, therefore, hard to prove. In a period of recession such as we are facing today, the TIC in most companies is subject to downsizing and even, in some cases, to justifying its very existence. It can be argued that, during recessions and the resultant corporate cut-backs, access to technical information becomes more, not less, vital to a company's well-being. Most senior managers do not think in this way. When they cut the budgets, the cuts are often uniform across the board — or worse — as R&D and their associate TICs suffer the biggest losses.

All is not gloom. Some TICs weather the economic storms intact and others, even in the most beleaguered companies, expand. It is this that I will focus on.

Images

From March through November 1992, Dr. G. Richter, head of Information and Documentation at OMV-AG in Austria, conducted a study of 25 TICs in the petroleum and chemical industries. The study was called IMAGES, which stands for Information Manager's Accomplishments & Great Expectations Survey. This was the second such study, the first one having been conducted a decade ago by another company. Since most questions in the two studies differed, a detailed comparison between the two is not possible. Of the 25 large organisations surveyed, 13 are in the United States, nine are in Europe and three are in Latin America; 14 are entirely privately owned and 11 are mostly or entirely government owned; 21 are companies and four are research institutes.

The study's purpose was to gather information on TIC staff, structure, policies and activities for use in demonstrating to companies the importance of having efficient TICs in a time of shrinking markets. Although the full study is available only to the 25 participating organisations, I am able to comment on its general conclusions.

Profile of the managers and staff

The titles of the 25 respondents vary from Manager to Supervisor to Head to Co-ordinator to Group Leader or Team Leader. Their departments are variously called Computer Facilities, Support Services, Information Systems, Information & Documentation, Library Services, Technical Information Centre, or Technical Communications. The majority (16) report to R&D divisions; the remaining nine report to administrative, economic or computer divisions. Only 10 work at the main organisation's headquarters office. Most (13) did search online and still do, but 12 never searched online. About 20 consider themselves to be information managers or information specialists, and the rest see themselves as librarians or systems analysts.

These managers on the average oversee a staff of 22, of whom 10 are professionals and 12 are non-professionals. Total staff size varies from two to 80, with the number of professionals varying from zero to 39 and the number of non-professionals varying from zero to 46. The ratio of professionals to non-professionals differs in each organisation; one company has 29 professionals and five non-professionals while another has 14 professionals and 46 non-professionals. As early as March 1992, staff reductions were anticipated. It was predicted that TIC staff size would diminish due to overall reduction, out-sourcing and decentralisation. Workload was already increasing and, with it, the need for more prioritisation. In general, staff size is deemed insufficient for the workload. The average employee works 41 hours per week. Enlarging the size of the staff is seen as very difficult by 21 respondents.

Salaries for TIC managers and staff keep pace with those in other corporate divisions. Most salary grades (14) are based on other units in the same organisation, but some are based on similar units in other organisations in the same industry or on local organisations in different industries.

Records management

Out of the 25 organisations surveyed, 23 have a records management programme, i.e., a central depository of corporate and/or research documents. The average size of the records management staff is six, with ranges from half an employee up to 30. In only 14 organisations is the records management programme a part of the TIC. In all organisations the records are indexed on an internal database. Access to this database varies widely for each organisation, from very restricted to almost public. Imaging technology is used at only four organisations. Lack of space is seen as a major problem for 16 organisations.

Data processing

The TIC and the data processing group are mostly separate units, though both tend to be in close physical proximity. In three organisations the DP group reports to the TIC Manager, and in three organisations the TIC Manager reports to the DP group. At most organisations (19) the DP group sets up company rules for hardware and software that the TICs have to follow. The amount and type of DP support vary. Most respondents are dissatisfied with DP support, which mainly exists in a mainframe environment. Half the respondents can easily acquire more hardware, and half cannot.

Online searching

Most respondents (19) expect their company's need for online searching to increase in the next five years, even as more databases, including full text, become available on CD-ROM. In general, CD-ROM products are still viewed very critically. Most respondents (18) prefer paying for connect time rather than having subscription-based pricing; the others can accept a combination of both. Cost is monitored when using different online hosts, but cost is not the most critical factor. Convenience comes first.

A wish-list of new features that respondents would like database publishers, online hosts or CD-ROM vendors to introduce during the next five years includes online graphics, online thesauri, daily updates, more full text, new types of output format, protocol downloading, similarity or associative searching, and seamless CD-ROM/online searching.

Although most of the organisations surveyed have end-user searching, the vast majority of the external online searching is performed by the TIC staff. More end-users search internal databases directly, but, even here, the TIC staff still does most of the searching. About half the respondents encourage end-user searching, provide training for it and expect their organisations to move toward more end-user searching. Most do not expect to install front-end software for end-users in the next two years.

Half the respondents offer information analysis and more are considering implementing it.

In-house databases

Most organisations (17) view the TIC staff as experts in developing and maintaining in-house databases and involve them in such. Most organisations (19) do their own abstracting and indexing, and this is increasing. On the average, eight employees per organisation perform this activity, with staff size ranging from one to 70. In only 12 organisations is the TIC responsible for this function.

Equipment/technology

The broadest possible range of equipment and software is cited by the respondents. No two TICs function in the same way regarding use of computer architecture, database software, library management systems, acquisitions systems, journal routing systems, archival technology, CD-ROM equipment, scanning technology, searching tools, LAN services and electronic mail. As for future technology, LAN and CD-ROM networking are seen as the big issues during the next five years.

Providing service

All the TICs serve R&D and almost all (21) serve Engineering. Most (14) also serve Strategic Planning, Economic Planning and Marketing. Only six serve Environment and Health, only three serve Legal and Patents and, curiously, only six serve Manufacturing. In theory, the number of clients served ranges from 600 to all employees, i.e., 90,000 (average: 18,750). In practice, the number of clients served ranges from 50 to 3,000 (average: 800). Most respondents (19) believe their client base is expanding. Only nine of the TICs charge back for services.

The number of other information centres in the organisation varies from zero to 70, with the average being 11. Most of the centres (16) are autonomous and the rest are centrally controlled. Some share subscriptions and engage in interlibrary loan. Most (15) share or have equal access to in-house databases. Some organisations are moving toward centralisation of information services and others toward decentralisation; there is no clear trend. The respondents are divided on whether centralisation or decentralisation is best. A few of the TICs (4) also handle business information, but this is the exception.

The crucial question of the survey asked: "Do you 'market' your information services throughout your organisation?" Only 19 (i.e., 76%) out of 25 answered 'yes'. The 19 affirmative respondents listed these marketing methods, with the asterisked (*) activities being cited most often:

- * onsite distribution of brochures,
- mailings,
- subscriptions list routing,
- use of questionnaires,
- customer interviews,
- * notices in company newsletters,
- notices on electronic bulletin boards,
- open houses,
- participation in information advisory terms,
- * introductory programmes for new employees,
- * presentations at the TICs, processing plants, regional offices and headquarters and
- personal visits to senior managers.

Conclusions from the study

Dr. Richter has noted two main conclusions. First, what he learned most from the survey was how to conduct a survey. This shows that, when dealing with complex issues, it may not be possible to construct a perfect questionnaire in one stage. Only after analysing the responses is it possible to come nearer to achieving the study's purpose, i.e., new questions must be formulated based on the initial responses. Second, although the study asked for responses based on the overall organisation, almost every respondent answered on the departmental level. This indicates a limit in vision and a constraint placed on serving the organisation as a whole. This constraint may be imposed by the organisation itself or, perhaps more often, it is self-imposed by the TIC manager.

More conclusions may also be drawn. First, it is possible to compare IMAGES I and II in one respect: stature of the TIC manager. In general, but especially in the

American companies surveyed, the TIC manager's position has diminished in the last decade. As managers have retired, some companies have sought to cut costs by abolishing the higher level position entirely or by reclassifying the job at a lower level. Other companies have used the retirement of the TIC manager as an opportunity to merge departments and jobs. These new TIC managers, with no background in information management, have found themselves in charge of a function they do not understand. The results are mixed. Some of the new managers have become fascinated with information technology and have used their contacts in other sectors of the company to widen the appreciation of information usage. Other new managers have failed to grasp the value of information and this may account for their timidity in marketing the TIC services and for the resultant shrinking profile of the TIC in their companies.

Second, most TICs are not located at headquarters and many are not adjacent to a major processing plant or refinery. To initiate ties with these sites, it may be necessary to travel.

Third, the study pinpoints at least six possible areas for most TICs to expand into: records management, information analysis, in-house abstracting and indexing, development and maintenance of in-house databases, liaison activities with other TICs and the Business Information Centres and, most importantly, increase in the client base beyond the R&D sector of the organisation. Serving mainly one sector is risky and violates the maxim: 'Don't put all your eggs into one basket'.

Beyond IMAGES

During the last five years, I have visited, at least once, all the 25 organisations represented in the study and have also visited about another 55 organisations in the petroleum and chemical industries worldwide. Based on these experiences, I have suggestions for internal marketing — new ones and amplification of suggestions listed above.

Ideas for internal marketing

The goal of internal marketing is for the TIC to serve the whole organisation to the maximum extent possible and to thereby increase the understanding of the TIC's value. To further this goal, it is necessary to:

- Broaden the definition of the TIC's mission
- Have more interaction with other corporate sites and subsidiaries, including foreign affiliates
- Become image conscious
  - Make TIC brochures look professional
  - Pay attention to the design and content of TIC announcements in in-house newsletters or on electronic mail
  - Improve the physical appearance of the TIC
  - Consider the appearance of the manager and staff
  - Present search results in attractive folders
- Increase the TIC's visibility by attending company functions and making contacts
- Make use of articulate members of the TIC staff
- Use customer feedback sheets and design them to give examples of TIC service benefits (cost avoidance)

- Collect, save and analyse these feedback sheets for an annual TIC report
- Follow up with non-respondents to the feedback sheets
- Know the TIC customers
  - Find out their areas of expertise
  - Find out how they work
  - Find out if they are the ultimate customer or the 'middle man'
  - Find out what kind of information they generally need
  - Offer them information before they ask
- Promote, do not resist, end user searching — if it did not work in the past, new technology (e.g., front-end software) may enable it to work in the future
- Provide the highest quality, most pleasant customer service possible
- Publicise the value of individual information sources to customers — this lets them know what the TIC (and they) cannot live without and it protects against budget cuts
- Offer more value-added services like information analysis
- Identify potential customers and offer them SDIs, e.g., news about their own organisation or competitive intelligence
- Stay attuned to what is important to the company overall

Throughout all this, it is important to keep statistics on workload. Otherwise it is impossible to justify expanding the staff size, should that become necessary.

Ideas for proving TIC value

For those TIC managers who can accomplish all the above and have time left over, there is still more that can be done:

- Have more interaction with TIC colleagues at other organisations
- Do not limit interaction to the same industry
- Exchange TIC promotional brochures
- Work collectively to create a programme for maximising information use
- Share ideas via electronic mail or by publishing articles
- Resist disempowerment by organisation and, especially, do not disempower yourself
- Travel more, not less, if possible — enlist colleagues to report on conferences when unable to attend; do this systematically
- Share information and power internally — secrecy and empire-building backfire in the long run
- Embrace new technology — be ready to make big changes and give up old systems (e.g., prepare for the coming of expert systems); develop a long-range plan
- Investigate out-sourcing before being asked — compare quality, confidentiality and long-term costs (it may be less expensive now, but ..)
- Find out if the company is, in ignorance, hiring consultants to supply information available less expensively through the TIC
- Ask for more funds and be prepared to justify them — the companies do have money and senior management must decide where to allocate it; managers are sensitive to upgrading the infrastructure

On the idea of intercompany co-operation through the use of surveys, M.H. Graham, formerly Manager of Information Systems & Services at Exxon Research & Engineering, has suggested that TIC managers design a questionnaire to accomplish four goals:

- prove to management the worth of the TIC to the organisation,
- protect against future downsizing,
- obtain quantitative measure of benefits and
- show how services can be expanded.

This is a big challenge, but, given hard work and creativity, it should be feasible. Frau Dr. U. Schoch-Grübler, Vice President of Documentation at BASF, has another challenge: for one company to share with other companies a case study of how a TIC saved its organisation from a needless major expenditure. She believes that only a real example will prove the worth of information to senior management.

Amplifying on the ideas of funding and also image, H. White, Professor at the Indiana University's library school, urges information specialists to refuse to provide service without resources. White believes that the spectacle of public librarians in Baltimore, MD and Pasadena, CA foregoing salary increases to compensate for acquisitions budget cuts was harmful to the information profession. "It suggests," he says, "that either librarians don't care about money, or librarians always find a way, or they get too much already." White has advice for seeking funding: create an awareness of what you are doing, then show what you can do with more funds and only then ask for money. "Create a situation where they can't live without you," he counsels. "You can't hide by being cheap. Cheap is not a strategy. Quality is the strategy." (As quoted in *Corporate Library Update,* August 1, 1993).

Dealing with senior management

The picture would not be complete without senior management, the group theoretically most able to appreciate the overall benefits accruing to the company from any activity. By finding allies among senior managers — even one ally — the TIC funding problems may lessen very rapidly. It takes even more work:

- Develop a special small brochure addressed just to senior managers
- Collect examples of how sectors other than R&D benefit from TICs
- Invite managers to visit the TIC and then follow up
- Focus on new managers and on how the TIC can help make their jobs easier
- Ask retiring contacts to introduce new contacts when possible
- Visit headquarters
- Learn senior management terminology — enhance the TIC's image by discussing it in these terms
- Avoid detail when addressing senior managers — speak or write in broad concepts, but have detail available to back up assertions
- Seek help of publishers and vendors to reach senior management
- Have presence at outside meetings attended by senior managers to learn of their concerns and to make contacts
- Develop a plan for expanding the client base upward
- Appeal to senior managers by developing a broader understanding of the company and finding ways for the TIC to benefit the whole company

- Find out how TICs function in other organisations and propose changes to emulate the superior ones
- Issue reports on major accomplishments showing how the TIC contributed to the bottom line
- Educate the TIC manager's immediate supervisor and turn this person into an ally
- If old contacts already exist, renew them — act fast before they retire

Conclusion

So much depends on the character of the TIC manager and staff. M.H. Graham sums it up well: "A new type of TIC manager is needed in the 1990s, one who is pro-active, who must have a broad view of the organisation, who must establish good relationships throughout the organisation, who must understand the need to establish value to the organisation for the TIC's services and who must look for opportunities in the very uncertain world."

One company that I have visited has a TIC manager whose first priority is protecting her job. When I first met her three years ago, the TIC staff size was 10. Worried about an upcoming corporate downsizing, she decided the best strategy was invisibility. She reasoned, "If they don't know I exist, they can't eliminate my position." She stopped travelling, stopped marketing the TIC services and kept a low profile in all ways. During the first round of budget cuts, her staff was reduced by three. She and her remaining staff were miserable as the work piled up and they could not keep pace. A year later, three more employees were let go. Today, everyone seems even more overworked and miserable, but the TIC manager still has her job.

Another company that I am familiar with underwent a major cut-back this year. The TIC survived intact. Its manager, as well as her predecessor, had laid the ground-work for years. This manager is very 'political.' She attends the corporate social functions, mingles with senior managers at every opportunity and looks and sounds like a senior manager herself.

The most startling success story I can think of comes from a very beleaguered company. During its huge downsizing last year, the TIC staff size increased. A few years ago, no company had a less appreciated TIC function. Even its immediate supervisor was openly hostile. A new TIC manager would not accept the lack of appreciation. She developed a long-range plan. After many battles, she turned her supervisor into a TIC supporter. She regularly travelled to headquarters and be-friended senior managers. When she heard that budget cuts were coming, she presented a plan to show how money could be saved by consolidating various functions and moving them into the TIC. Senior management listened.

The survival of the TIC is up to the manager and the staff. The success of the TIC is not directly related to its size or to the company's size. It is directly related to limits that people create for themselves. Ultimately, survival is not enough and it does not do justice to the whole organisation. The motto for the TIC of the 1990s should be: Don't just survive; *thrive*.

Searching chemical information in the EPO: an overview of current methodologies and a discussion of future developments

John Brennan

European Patent Office, DG 1, Rijswijk, The Netherlands

Introduction

In the EPO Search Division, DG1, there are approximately 300 search examiners working in the technical areas covered by the broad term chemistry, which extends from such inorganic areas as cement and alloys through organic fields such as foods, polymers and pharmaceuticals to the biological areas of *in vitro* testing methods and genetic engineering.

Given this diversity of technologies it is perhaps not surprising that providing a short coherent description of how a search is carried out in chemistry at the EPO poses the same problems as asking for a description of how a chemical substance is produced or how chemical research is carried out; whether we are talking about search at the EPO or industrial manufacture or research, only general approaches and overall organisational structures are shared, while the specific methods and tools will depend upon whether it is cements or monoclonal antibodies which are of interest.

The objective of this presentation is to give an overview of the general approaches and organisational structures used in the EPO and to go into further detail to exemplify the methods used in some specific technical areas. In this respect I shall adopt a chronological approach moving from the paper led approach which was used in the beginning of the EPO in 1977, to the online led approaches which are currently being developed, and finally to how things may be seen as evolving using an online-only approach.

Firstly however, in order to provide a clear background to what will be dealt with later it will be necessary first of all to describe how the examiners and documentation are organised.

1) Organisation of examiners

In The Hague there are 11 search Directorates in chemistry, each composed of between 20 and 30 examiners and a director. Each directorate examines applications in a group of related International Patent Classification (IPC) subclasses; for example one directorate examines all applications for new aliphatic, alicyclic, and heterocyclic compounds (IPC C07C and C07D), for all uses; within that field an individual examiner will usually examine only applications with their principal classification in a limited number of groups, for example heterocyclic six-membered rings containing only one nitrogen atom (IPC C07D211-C07D221). Clearly, because of the way in which chemical claims are usually formulated, with variable functional features, an examiner will often have to search outside such a core area, but because of this method of organisation each examiner should have a high level of familiarity with developments in that core area.

2) Organisation of the documentation

The paper search documentation consists of so-called 'search groups' each of which comprises patents and possibly journal articles and abstracts classified according to the European Classification (ECLA); ECLA is mainly based upon the IPC, with finer sub-divisions of the IPC sub-groups where examiners have felt that such sub-division was desirable. A readily comprehensible example of this internal sub-division which was formerly used in the area of monoclonal antibodies (and has now been further developed) is shown below.

C12P21/08	. monoclonal antibodies
21/08A	..[N: against material from animals, e.g. liver fluke, schistosoma]
21/08A1	...[N: against B- or T-cell antigens]
21/08A2	...[N: against blood group antigens]
21/08A3	...[N: against HLA antigens]
21/08A4	...[N: against blood coagulation factors]
21/08A5	...[N: against tumor antigens]
21/08A5B	....[N: of breast]
21/08A5C	....[N: of lung]
21/08A5D	....[N: of liver or pancreas]
21/08A5E	....[N: of kidney, bladder or prostate]
21/08A5F	....[N: of stomach or intestine]
21/08A5G	....[N: of skin]
21/08A5H	....[N: of blood cells]
21/08A5K	....[N: against oncogenic proteins, e.g. RAS encoded proteins]
21/08B	..[N: against material from plants]

In this case the IPC only provided one sub-group for all documents relating to monoclonal antibodies, C12P21/08; this was divided in ECLA into antibodies against animal material and antibodies against plant material. Because most documents related to animal antigens, the animal part was further divided according to the types of animal antigens, and because most documents relating to animal antigens were specifically concerned with tumour antigens, this part was further sub-divided according to the type of tumour. The letter directly after the IPC class indicates that this is an ECLA sub-division and the symbol N: in the text indicates that the text is ECLA.

The purpose of such sub-classification is to try to keep the number of documents which are to be consulted during a search more or less constant as the number of publications in a broad area of technology increases. This is important in trying to maintain productivity as the total documentation grows. At the same time however it is important that such divisions of the search groups are clearly structured and allow many searches to be made in a reduced number of documents with no loss of relevant information. That is, that the divisions reflect general and probably continuing trends of development within a field of technology which have become separate sub-fields, usually independent of the other sub-fields. In this respect the examiner's expert knowledge of current developments in the field is of key importance in making appropriate divisions.

The documents present in the 'search groups' arrive there as a result of their classification by EPO examiners; examiners classify only in the fields of their own expertise, and documents are circulated by and amongst examiners working in all possibly relevant fields and wherever appropriate are multiply classified. The documents can be divided into three broad types:

- European patent applications (EP's): these are studied and classified by the search examiners, normally during the search, or prior to their publication as A2 documents.

- Non-European patent applications: where there is no EP family member the non-EP documents are studied and classified by the search examiners taking into account any classification assigned by other patent offices. Only the first published family member from the PCT minimum documentation is included in the classified documentation; where there is no family member in the PCT minimum documentation Belgian, Dutch, Luxembourg and Swedish documents are also included.

- Non-patent documents: in certain fields abstracts, journals and other sources of technical information are systematically reviewed by examiners, and relevant parts are classified and photocopied.

As far as is reasonably possible, all of the claimed subject matter is classified (that is, a copy of the document is placed in each appropriate 'search group'). For broad claims in chemistry, especially of the Markush type, it is often impractical to follow this principle to its ultimate conclusion; for instance, a document comprising a claim for new compounds with a variable such as "R2 is an optionally substituted aromatic ring which may be carbocyclic or may contain from 1 to 3 nitrogen atoms or an oxygen or sulphur atom, or a combination of these" would fall into hundreds of classification sub-groups. In such cases, a balance must be struck between including what might be regarded as technically or legally significant, and avoiding dilution of the documentation with documents of marginal significance with respect to technical disclosure or potential legal effect.

Classical paper-led searching

From 1977 when searching began in the EPO, it was carried out in generally the same manner as had been practised in most other patent offices for the past century. An application was studied, appropriate classes were designated and the 'search groups' corresponding to these classes were located; the documents present in the 'search groups' (inherited from the former IIB and the Dutch Patent Office) were examined as were, for example, the paper texts of Chemical Abstracts, Agdoc and Plasdoc where appropriate. This part of the search was directed towards both novelty and inventive step and any documents relevant to either of these aspects of the application were cited in the search report.

Even at this early stage however, mechanised searching was being applied in areas such as laminates (IPC B32B) and lubricants (IPC C10M) where the use of IPC groups and sub-groups, or a related ECLA was considered as being ineffective. In such cases all of the relevant full-text patent documentation was stored in country/numerical sequence; codes, punch cards and pins were used to generate a list of documents relevant to each individual search, which were then manually retrieved from the collection, and used as the basis for the search.

As online sources became available these were used to supplement the paper search, for example Chemical Abstracts bibliographic searching began to be used in 1978, Agdoc has been used in the pesticides and herbicides area (IPC A01N) since 1979 and structure searching in DARC commenced shortly after it became available in 1981. As a result of this, by the mid 1980's examiners in the field of pure organic chemistry (claims for new low molecular weight compounds) had abandoned the systematic classification of photocopied journal articles, on the basis that most significant non-patent information could be retrieved online.

Until the mid-1980s however, the online element of most chemistry searches was secondary to that of the paper and in some cases totally absent, and structure searching, when it was carried out, was heavily directed towards novelty. This limited use of online searching was to some degree a consequence of the fact that most of the online interrogation was carried out by 'operators' or 'intermediaries' who were provided with search terms or substructures by the examiners. Clearly, while online searching retained the mystique of a highly specialised activity which could only be carried out by a select group of experts, it could never play more than a secondary role in the overall searching process.

In summary therefore, in most cases the search with respect to both novelty and inventive step began with and was principally based upon direct inspection of the classified paper documentation, and online was used as a tool for filling in any subsequent 'gaps'.

Online-led searching

Around 1986 the move towards online searching being carried out by the end-users commenced, and from 1985 to 1990 the percentage of examiners in DG1 executing independent online searches increased from around 20% to around 95% (see Graph below); it is reasonable to assume that in chemistry the latter figure would now be close to 100%.

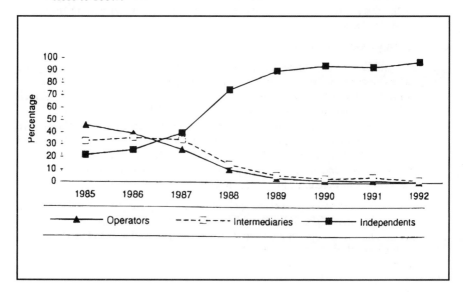

At the same time as most examiners were becoming independent searchers of external online systems, all examiners were being provided with pc's connected by a token ring local area network to the mainframe computer; this was to allow examiners to be able to take advantage of a number of important developments which were taking place with the internal computer documentation systems.

1) Epoque

This is an internal host system available to all examiners which comprises 36 databases loaded on our mainframe (including the integrated WPI/L (WPI), Patent Abstracts of Japan (PAJ) and the former mechanised systems) and allows access to all major external hosts. A more complete description of all of the planned features of Epoque was given by Annemie Nuyts [1] from the EPO at the Montreux meeting in 1989 and an update on the operational system was given at the 1991 EUSIDIC meeting in Seville [2]. The main screen of Epoque is shown below.

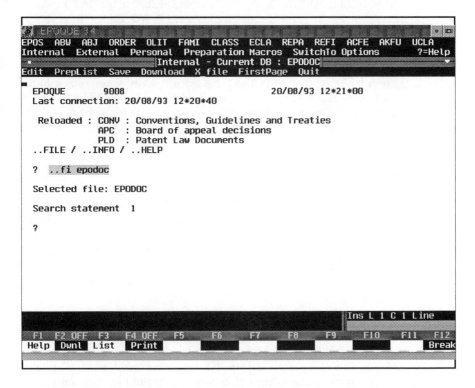

In summary, as well as providing searchable databases with the possibilities for left-hand truncation, cluster searching, cross-file searching and the automatic uploading of queries, it allows the manipulation, downloading and printing of results (including chemical structures); furthermore, the 'First Page' function allows the visualisation of drawings and formulae present on the first pages of documents retrieved (including JP abstracts from PAJ).

While in areas such as laminates the first step in searching has always been automated, the potential for manipulating and interrelating information offered by

Epoque clearly offers the opportunity for examiners to explore alternatives to the classical paper-led searching in many other areas. Much of what follows therefore is a description of techniques which may be recently introduced methodologies in some fields or experimental approaches in others.

2) Internal Databases

At a simple level, instead of an examiner going initially to the appropriate paper search group in order to see the application being searched in context of the relevant prior art, the corresponding approach may be taken online using the EPODOC database in combination with WPI and PAJ.

2.1 Epodoc

EPODOC comprises comprehensive bibliographic data on all of the documents present in the classified collections and their family members; all of this data is available in EPOQUE, including a searchable ECLA index, and key-word searchable English language abstracts and titles of EP, WO, US or GB documents in the basic index.

It is possible to use the Epoque software to memorise the priority numbers of all of the patent documents present in a specific search group in EPODOC; after going to a cluster of WPI, PAJ and EPODOC, the list of records corresponding to the priorities can be generated from the memory, and this can be searched by combination with appropriate terms relating to the application. At this point the records corresponding to these hits can be viewed (with the possibility of using the formulae or drawings in First Page as an additional filter), and a decision can then be made as to how to proceed with the search. This approach has the advantage over the paper that highly relevant documents within a 'search group' are identified immediately; within the numerical ordering of a given search group the technical information may be more or less randomly arranged and considerable time can be wasted during a paper search studying documents which cease to be of importance in the light of the content of those arranged subsequently.

Because the ECLA class has been used as a filter to define the group of documents being searched, broader key-words can be used for limiting the answer list compared with a simple key word search, and the number of false drops can be reduced.

For example:

In a search for halogenated derivatives of glycosylated macrolides with 20 or more ring members the search can be carried out in the ECLA group C07H17/08G; transfer of the priorities corresponding to this group from EPODOC to the WPI/PAJ/EPODOC cluster gives rise to 255 records and combination with general and specific halogen search terms gives 9 documents; in First Page images are available for 3 relevant documents and 2 false drops; the remaining 4 records do not have images but can be clearly identified as false drops from the text. For the three relevant documents the patent and non-patent citations from their search reports are available in EPODOC.

By comparison, carrying out the same search only in WPI and using C07H17/08 as the IPC search term in the IC field gives 315 answers. It is also worth noting that such a general search could not be easily executed as a substructure search in a structural database.

A related approach to searching using EPODOC and Topfrag-generated Derwent fragment coding is currently being explored. Until now application of the fragment coding as a search tool has only been used by a minority of examiners, working in fields where frequent use has allowed familiarity with the rules for coding; in other areas searching of the classified paper (possibly in combination with a CA structural search) has been regarded as providing results of at least an equal quality. The approach now being assessed consists of the transfer of the patent priority numbers corresponding to the contents of a 'search group' into WPI to generate the document list, and then to use the Derwent fragment coding to define structural features which are not taken account of in the classification and to use their combination to limit the answer lists.

At the moment this is being carried out by some examiners in pure chemistry who can use the coding; however, with a view to the more general use of this approach, structures are being created in Generic Topfrag, modified to remove unwanted restrictions, and automatically transferred into the Epoque WPI session. This work is in anticipation of the potential use of the Markush Topfrag software, which is intended to automatically remove codes which would be in conflict with the presence of 'free sites'; as opposed to the sole use of fragment coding, combination with EPODOC should significantly reduce the number of 'false drops'; false drops are an accepted feature of completely defined coding, and will clearly become more numerous with the use of a less exhaustive coding in the Markush version. At the same time, the expected relative lack of complexity in creating correct coding using a software such as Markush Topfrag and the ability (using software already internally available) to present the output in a form directly suitable for automatic searching in Epoque should make this approach accessible to all examiners working in areas where fragment coding is present in the relevant Derwent records. In principle this should even extend to examiners who only occasionally have need of such an approach.

While such EPODOC-ECLA based online searching, using key-words in WPI, PAJ and EPODOC or fragment codes in WPI, ultimately relies upon the information content of abstracts (as opposed to full-text), this limitation may in many cases be at least balanced by the possibility of carrying out the search using several distinct strategies and in a much larger number of possibly relevant classification units than would be practicable using paper. It is also clear that the long term need for high quality classification and the continuing introduction of new sub divisions within ECLA is in no way diminished.

2.2 Epos

In order to allow thorough key-word interrogation (especially of limited data sets, such as the EPODOC generated lists corresponding to the paper search groups), the EPO is currently building up its own synonym thesaurus, named EPOS. The EPOS records are provided by examiners for search terms which frequently occur within their fields of specialisation and this offers a number of potential advantages:

i) consistent use of a comprehensive set of synonyms with correct spelling in bibliographic searching for examiners normally working in the field;

ii) availability of the set to examiners who may have to search aspects of applications outside the area of their normal expertise;

iii) an additional linguistic support for examiners who do not have English as their first language.

While the use of EPOS originated in the non-chemical fields, records are now being created in the pharmaceutical and medical fields (e.g. synonyms for specific diseases and biomechanistic modes of action).

2.3 Additional computer coding

As well as the former mechanised systems which employed their own coding, coding systems based upon the IPC and ECLA and designated CIS (complementary information system) and ICO (in computer only) have been applied to documents in a variety of areas including 13 chemical fields; these codes may relate to supplementary or complementary features of an invention. This information is stored on the mainframe with the other data for the document and is available and searchable through Epoque. The principle here is that the time spent in systematic coding of such features is soon compensated for by minimising the need to re-read the same lists of alternatives in connection with many subsequent searches.

In the area of surgical articles the ECLA class A61L31/00E5 relates to polymeric material other than addition polymers for certain surgical articles; this may be linked in the CIS database to the (more specific) class for polyesters derived from hydroxycarboxylic acids, C08L67/04, for example:

/c A61L31/00E5 L C08L67/04

This gives an answer set of 51 in-file documents. Use of only A61L31/00E5 in EPODOC gives rise to 124 documents, whereas combination of this set with C08L67/04 in the WPI classification (IC) index retrieves only 9 records. Use of the coding has therefore allowed the detection of five times more documents, which explicitly disclose the desired combination of features, than a simpler approach, but still less than half of the number from use of only the main ECLA class.

In the above example the linked combination has been made between features from two distinct sections of the IPC (A and C); coding can alternatively be used to indicate relationships between different IPC groups within the same subclass. Where synergistic compositions of a specific group of herbicides with various other alternative types of herbicides are disclosed, the classification relating to the first group may be linked to the classes corresponding to the alternatives. This takes place at two levels; appropriate A01N classification is applied to named herbicides in mixtures and furthermore an indicator is applied to invariant components. For example, in a search for herbicidal mixtures of a 1,3-diazine (A01N43/54) with any class of organic phosphorus herbicide (A01N57), the online search statement would take the form:

/c (A01N43/54% L A01N57/??) OR (A01N57/??% L A01N43/54)

Here the % indicates that this class must be coded as an obligatory feature linked by the L operator to any sub-group of A01N57, where ?? are optional characters. This coding therefore adds information present in the description and at the same time may be used to filter out 'noise' arising from the presence of classes which correspond to alternatives to each other. In the present case 15 documents were retrieved as opposed to 46 using straightforward classification combinations in WPI.

The above two examples illustrate how, in the first case coding allows the retrieval of additional documents where two distinct features of the same invention may be disclosed but one is not claimed, and in the second case where coding can be used to distinguish between similar features being present as alternatives or obligatory combinations in the same document.

The ICO system which is also used in certain areas differs in that the ICO codes are not IPC groups; for example searching in EPODOC:

C12N9/64C/ec AND M12N207/00/ico

This search statement relates to a combination of the ECLA sub-group (/ec) for Factor IX proteinase derived from animal tissue, and the ICO group for enzymes prepared by recombinant DNA technology. It is worth noting that searching only at the IPC group level (C12N9/64) gives 294 in-file documents, limiting this to the ECLA sub-group (C12N9/64C) gives 39 documents and combination with the ICO code limits the answer group to 18 documents.

The uses of coding by examiners expert in its application in their fields may of course be much more varied and sophisticated than in the above examples which were chosen for their relative simplicity and ready comprehensibility.

3) Personal databases

The facility exists within Epoque for examiners to create their own field searchable databases which also comprise such possibilities as crossfile searching. They contain data (key-words, text and citations) relating to searches which examiners have carried out personally; because of the limited technical areas in which examiners work, a volume of relevant data can soon be amassed, which provides a useful and flexible supplementary search tool. The potential also exists for such data to be combined, and in the photographic area such a combination has recently resulted in the creation of the generally accessible PHOTO internal database.

4) External databases

As well as much greater use being made of internal online searching, more exhaustive searching through external hosts is also being undertaken. Lower connect time costs and greater expertise of examiners in strategies for the manipulation of search terms and results have allowed the online element of searches to increase significantly at acceptable costs.

4.1 Substructure searching

The use of substructure searching in EURECAS and REGISTRY has also been considerably developed. Even for novel compounds, this part of the search now frequently commences with a broad query intended to generate hundreds or even thousands of structures which after the transfer of Registry Numbers can be used as the basis for a bibliographic search directed towards inventive step.

A much narrower Boolean query based on the broad structure set is used as the first stage of the novelty search, followed by the use of the broad set to generate a list consisting of patent documents not so far identified. The last part of this strategy is intended to take into account the Markush factor, on the basis of the fact that disclosures in non-patent literature do not normally extend beyond the scope of what has been exemplified whereas patents do, and so the scope of Markush claims of patents exemplifying structurally similar compounds to those being searched should be checked.

The effective application of this last part depends upon the ready local access of patent documents, possibly from a diversity of fields; at present, the classified paper documentation satisfies this need to a reasonable extent and in future this might be more efficiently satisfied by rapid electronic access to the images and formulae of the claims or examples in documents retrieved from numerical collections.

While the remarks above have been directed mainly to the area of pure chemistry, evaluation of both the Topfrag and CAS-structural approaches in the applied area of photographic chemistry is being undertaken. Here the lists of hits will almost always be large; in WPI combination with EPODOC classes will often provide acceptable document lists; while CAS sub-structural searching may give rise to lists of thousands of compounds present in tens of thousands of documents, often only a small percentage of these will have been used in photographic processes and these can be retrieved by appropriate use of a section code and/or key words. Again, ready access to the content (including chemical structures) of patent documents is then essential.

4.2 Markush databases

In his paper delivered at the 1991 Montreux conference, I was quoted by Jim Sibley (3) as having stated during the 1991 EPO DG1 Seminar for Applicants that no significant use was being made by the EPO of the Markush databases; the reasons given by me for this (limited time range, general adequacy of the CAS and Beilstein structural databases and low value of peripheral documents retrieved) were reiterated. Much of the argumentation presented by Everett Brenner and Edlyn Simmons (4) at the 1992 Montreux meeting would also be considered by me as pertinent. At the same time however, at the EPO we continue to monitor developments in the Markush databases with regard to our internal documentation, in order to assess the potential application of these databases in areas of rapidly increasing documentation where other solutions do not exist, or where such solutions would only be replicating the work already carried out by the database indexers in order to achieve essentially the same results.

As an example of this, testing is currently being carried out on applications classified in the parts of the IPC group C07H19 which cover nucleosides and nucleotides; this includes documents relating to the anti-AIDS compound AZT and many similar structures. The area is considered as being one possibly appropriate for use of the Markush databases for the following reasons:

i) A large expansion in this field has taken place since 1987 which makes the limited time range coverage of the Markush databases more acceptable. (The AZT applications which provided much of the stimulus for the expansion in this area were published in late 1986, and a substantial volume of the patent documents in the relevant parts of C07H19 have been published since the beginning of 1987.)

ii) Many of the complex Markush claims and corresponding examples in this area are formulated in such a way as to require searching in many sub-groups.

iii) Even where the number of relevant sub-groups is small, the number of documents present in these is often large.

iv) The stage is being reached where, if the high rate of growth in applications in this area continues, reorganisation or other alternatives must be considered.

v) A useful method of ECLA sub-division is not apparent, and internal coding would require a large amount of effort with respect to both retrospective and continuing work.

It is at present too early to see whether or not Markush databases will provide a solution in this particular area; it is however an example of the type of situation where their use is being actively explored, and hopefully this will dispel any impression which may have arisen, that the EPO has set itself in principle against the use of searching in Markush databases.

After the use of one or several of the above online approaches, it is at present normal for the examiner to then go to parts of the paper documentation which could remain significant in order to complete the search.

Moving towards online-only searching

From what has been so far described it is apparent that examiners now have many more electronic tools available than in the past. Clearly time spent in their use must be offset by a reduction in time spent in the paper documentation and equally this must be justified by advantages over use of the paper.

In many chemical fields the use of online methods to cover patent documentation (JP, SU, HU etc.) beyond the scope of the classified collections and non-patent literature is already standard practice because of limitations of language and/or breadth of coverage. The fact that the online searches also recover significant numbers of documents present in the classified collections must lead to questioning of the duplication of effort.

Because inventions sometimes do, and sometimes do not fall within one or a few classification units, searches in the paper often have to be limited for reasons of time and effort to the most relevant sub-groups; this can be a significant limitation on the comprehensiveness of the novelty part of the search and a more severe limitation on the inventive step part; online methods allow the indirect searching of many more classification units than would be manually possible, or allow the search to be carried out independently of classification.

Within a reasonably sized ECLA sub-group (say 150 documents), often 80% of the documents present will be of no real relevance.

It is also a fact of life that the amount of paper documentation present in the classified collections is increasing at a rate of around 5% per year. Even with effective ECLA sub-classification and reorganisation (which itself involves diversion of examiner search time) the number of documents which will have to be inspected during a search will increase. Since it is foreseen that the number of searches per examiner per year should not decrease and that search quality should be maintained, electronic means appear to offer the best method of identifying relevant documents and so limiting the number of documents which have to be physically checked by the examiner.

1) Assessment of online possibilities and limitations

The progression from paper-led searching to online-led searching gives rise to questioning the degree to which the online part completely and effectively replaces the paper, and which aspects of the paper (or images of the paper displayed electronically) would remain significant or indispensable parts of the search. In order to obtain some quantitative answers testing has been carried out on applications in the IPC sub-class C07H (carbohydrates). A similar test is currently under way in the applied field of photographic chemistry. This consists of the following steps:

i) The application is studied and a note is made as to whether the application is considered to be one for which an online search alone would be expected to give good results, or if online in combination with a search of the pre-1970 documentation would be expected to give good results. The year 1970 is used on the basis of the introduction of comprehensive WPI coverage in that year; for searches in the Pharmdoc (1963) and Agdoc (1965) areas, combination with searching in the CAS structural file (1967) could make 1967 an appropriate cut-off.

ii) Exhaustive online searching using all approaches which might be considered relevant (ECLA sub-group and broad key-words, ECLA group and narrower key-words, IPC and key words, external substructure and bibliographic etc.) is carried out and the results are combined.

iii) The patent documents are then removed from the corresponding classified 'search groups', documents from the non-patent literature where internally available are photocopied, or if not available are ordered from external sources (again via Epoque).

iv) Where pre-1970 documentation was considered relevant the search is extended to the paper documentation in that time range and relevant documents are selected.

v) The search is then extended to all relevant classified paper documentation not so far searched, and any further significant documents are selected.

vi) Where existing, the online records corresponding to the documents retrieved under point 5) are examined and the reasons for the failure to retrieve them are assessed.

The overall coverage of documentation using this approach is the same as would be used in a conventional approach but documents which would be missed in an online-only approach are clearly identified.

The preliminary conclusions are as follows:

• That it is normally possible for an experienced examiner to judge the time range in which the retrieval of relevant documents is likely.

• Where the post 1969 documentation was comprehensive with respect to the technical field of the application being searched, the following points can be made with regard to online searching:

i) When lacking experience in this approach, the initial tendency of the searcher is to try to narrow the online search to retrieve only relevant documents for inspection of the full-text and this results in loss of other relevant documents (especially patents). Experience showed that a broader query strategy, which resulted in a significant number of false drops only detectable at the stage of full-text consultation, was necessary to retrieve all relevant documents.

ii) That even for applications directed primarily towards novel compounds a relevant to false drop ratio of around 1:1 has to be accepted to allow the identification of all relevant full-text patent documents. Given that the search report consists of a further selection of the most relevant of the documents inspected in the full-text, between 10 and 20 documents may have to be inspected in the full-text form.

iii) For applications directed towards processes for bulk chemicals (e.g.fatty-alkyl glycoside surfactants) a similar degree of redundancy of around 1:1 often has to be accepted for relevant documents, but the list is normally much larger than for new compounds because the online means are not available for restricting the list to those documents relevant to the significant process parameters of the application. In such cases, between 20 and 50 documents may have to be inspected in the full-text form.

Indicating that documents have to be inspected in the full-text form is not to say that the whole document has to be read; it may often be the case that documents may be quickly excluded on the basis of inspection of only part of the full-text (especially claims, formulae, drawings, examples or a small part of the description).

Consequently, in many cases in pure chemistry, exhaustive online searching in Epoque, in combination with an electronic server capable of delivering the images of 20-30 patent documents would be acceptable for the execution of a good quality search. For processes, possible solutions could exist in coding of the in-file documents in order to allow the generation of a smaller document list for full-text image inspection, or in the intermediacy of full-text searching, possibly in combination with the use of EPOS, between the online list and the full-text images.

In applied chemistry (e.g., photographic chemistry) the situation would appear to be similar to that for processes in the area of pure chemistry; in the absence of additional filters the number of documents to be inspected in the full-text image format after online searching can often be in the region of 50; as well as presenting electronic handling problems, the on-screen inspection of such a number of (often lengthy) documents may be unacceptable to the examiner. The use of coding or intermediate full-text searching are also the most obvious solutions to be considered here.

As above, where the pre-1970 (or pre-1967) literature contains relevant documents, online searching in the conventional databases clearly does not offer a solution. In many organic classes they represent around one third of the in-file patent documents. Potential solutions to the problem of this time range include:

i) availability of the pre-1970 paper for manual searching;

ii) visualisation of the full-text of all of the pre-1970 documents (because pre-1970 patent documents were generally much shorter, the processing capacity of the system could allow the handling of many more documents than the 20-30 corresponding to an up-to-date average sample);

iii) the application of coding only in the relevant time range and the use of EPODOC to generate lists of time and code restricted documents for full-text visualisation;

iv) use of OCR to digitise text in areas where (in the light of experience in fields with available digitised text) full-text searching is an effective tool.

From what has been said here it is clear that what is presently being done in the online-led approach to searching provides a route for developing the strategies which will become fundamental in the increasing reliance upon electronic technologies. The subsequent searching in the paper provides a feedback as to where weaknesses were present in the online strategies, and in areas where attempts are now being made to increase the reliance upon online the systematic approach of checking the records of documents subsequently retrieved in the paper allows the modification and optimisation of online strategies, with no loss in quality in the searches currently being carried out. When subsequent consultation of the paper documentation in a technical area consistently ceases to provide anything better than what has been found online, then it seems reasonable to abandon the paper search in that area.

2) Applicants' reservations

The apprehension is felt by some applicants that we are sacrificing a search tool (the classified paper collections) which sometimes provides them in the search report with documents which they did not themselves retrieve, for a tool to which they have equal access (online) and from which they can already obtain the same results as us. This seems unjustified for a number of reasons.

As indicated in the section on the assessment of paperless searching, our objective is to develop online strategies to the point where the retrieval is effectively the same as by searching in paper.

Epoque facilities go far beyond those available for simple online searching using a PAD. In the same way that our classified documentation has been constructed to optimise the use of paper, Epoque allows the rapid searching, interrelation, combination and filtering of electronic information in such a way that the volume of data which can be processed in the time available for carrying out a search vastly exceeds that which could be handled using conventional online techniques. This will be combined with the facility of having immediately available the full-text form of any potentially relevant documents, which minimises the chance of a wrong decision being taken on the basis of an abstract.

With regard to our paper collections and the general access of applicants to online hosts the following observations are perhaps relevant:

- Information on the content of our classified paper collections has been available to the public for a number of years. It would seem surprising for a company with a large financial commitment in a specific technical area not to have obtained any potentially useful documents which are available to us, and are not already present in their internal collections.

- Claims in applications, even from large companies and even for compounds *per se*, often contain subject matter which is found to lack novelty on the basis of substructure searching in the commonly available hosts. The fact that applicants can directly find such information is no guarantee that they do, or if they do, that they will take account of it in drafting their claims.

- In the case of electronic information, as with the paper, the question is not whether applicants could in time and at some cost come to the same conclusion as a searcher in the EPO, but whether an applicant would in normal circumstances carry out all of the online searches, display all of the results, and order in full-text copies of all of the possibly relevant documents in order to come to that same result.

Conclusion

Searching in the EPO in most chemical areas already extensively uses online methods. These are currently being adapted and assessed with regard to their playing a primary role, in combination with technologies currently under development, in retrieving the documents cited in the European Search Report. During this phase, the continuing presence of the classified paper provides a safeguard, and a means of checking the validity of the new methods. This gradual approach also allows examiners to progressively develop their skills in the optimal use of the new technologies in the same experiential way as they developed them when they began to carry out searches in the paper documentation.

It therefore seems reasonable to conclude that the change to computer based searching in chemistry will allow the office at least to maintain the quality of its searches and the productivity of its examiners in the face of the significant annual growth in the volume of technical information which must be processed.

Citations

[1] A. Nuyts, 'Present and Future EPO Systems for Automation of the Search in Directorate General 1 : EPOQUE, BACON and CAESAR', in Proceedings of the Montreux International Chemical Information Conference, H. Collier, Ed., Springer-Verlag, 1989, p.187-190.

[2] A. Nuyts, in Proceedings of the EUSIDIC Annual Conference, Seville 1991.

[3] J. Sibley, 'If only...a sideways look at some patent databases', in Proceedings of the Montreux International Chemical Information Conference, H. Collier, Ed., Royal Society of Chemistry, 1991, p.21-32

[4] E. Brenner and E. S. Symmons, 'A Markush Story', in Proceedings of the Montreux International Chemical Information Conference, H. Collier, Ed., Royal Society of Chemistry, 1992, p.127-145

Index